SpringerBriefs in Economics

AF774335

For further volumes:
http://www.springer.com/series/8876

Carol Yeh-Yun Lin • Leif Edvinsson
Jeffrey Chen • Tord Beding

National Intellectual Capital and the Financial Crisis in China, Hong Kong, Singapore, and Taiwan

Springer

Carol Yeh-Yun Lin
Department of Business Administration
National Chengchi University
Taipei, Taiwan

Jeffrey Chen
Accenture
Chicago, IL USA

Leif Edvinsson
Universal Networking Intellectual Capital
Norrtälje, Sweden

Tord Beding
TC-Growth AB
Gothenburg, Sweden

ISSN 2191-5504 ISSN 2191-5512 (electronic)
ISBN 978-1-4614-5983-5 ISBN 978-1-4614-5984-2 (eBook)
DOI 10.1007/978-1-4614-5984-2
Springer New York Heidelberg Dordrecht London

Library of Congress Control Number: 2012951340

© The Author(s) 2013
This work is subject to copyright. All rights are reserved by the Publisher, whether the whole or part of the material is concerned, specifically the rights of translation, reprinting, reuse of illustrations, recitation, broadcasting, reproduction on microfilms or in any other physical way, and transmission or information storage and retrieval, electronic adaptation, computer software, or by similar or dissimilar methodology now known or hereafter developed. Exempted from this legal reservation are brief excerpts in connection with reviews or scholarly analysis or material supplied specifically for the purpose of being entered and executed on a computer system, for exclusive use by the purchaser of the work. Duplication of this publication or parts thereof is permitted only under the provisions of the Copyright Law of the Publisher's location, in its current version, and permission for use must always be obtained from Springer. Permissions for use may be obtained through RightsLink at the Copyright Clearance Center. Violations are liable to prosecution under the respective Copyright Law.
The use of general descriptive names, registered names, trademarks, service marks, etc. in this publication does not imply, even in the absence of a specific statement, that such names are exempt from the relevant protective laws and regulations and therefore free for general use.
While the advice and information in this book are believed to be true and accurate at the date of publication, neither the authors nor the editors nor the publisher can accept any legal responsibility for any errors or omissions that may be made. The publisher makes no warranty, express or implied, with respect to the material contained herein.

Printed on acid-free paper

Springer is part of Springer Science+Business Media (www.springer.com)

Foreword 1

The economic crisis is a consequence of many parallel factors which are all related to globalization and digitalization. My main concern, assessing this in more detail from the European perspective, is that revolutionary global forces have not been taken early nor seriously enough by most national and regional decision makers. The heads of European States and governments have once again recalled the importance of fiscal consolidation, structural reform, and targeted investment to put Europe back on the path of smart, sustainable, and inclusive growth. The main question is how capable and ready are the national governments to tackling the complex and manifold issues of crises and to renewing even radically many of our public and private structures and processes.

The first basic requirement is that all the European Union Member States remain fully committed to taking the actions required at the national level to achieve the objectives of the Europe 2020 Strategy. The second basic requirement is that the national and regional governments, as well as people, are ready for radical changes. This booklet and the other 11 booklets by the experienced authors focus on national intellectual capital and give necessary insights and facts for us the readers and especially for our in-depth systemic thinking of the interrelationships of NIC and economic recovery.

How should the national and regional decision makers tackle the existing knowledge of intangible capital? The focus needs to be more on the bottom-up approach stressing the developments on local and regional levels. I highlight our recent statements by the EU Committee of the Regions. The key priorities are to get more innovations out of research and to encourage mind-set change toward open innovation.

The political decision makers are finally aware that the traditional indicators created for and used in industrial production cannot be applied to a knowledge-intensive, turbulent, and innovativeness-based global enterprise environment. Indicators that perceive the intangible dimensions of competitiveness—knowledge capital, innovation knowledge, and anticipation of the future—have been developed around the world, but their use has not yet become established in practice. This booklet accelerates the development and the use of these indicators.

This helps the local and regional, as well as central, governments in taking brave leaps forward on a practical level—giving greater ownership and involving all the stakeholders. This means the need of actions toward increasing the structural and relational capital of regions, both internally in communities of practice and in collaboration with others.

The new generation innovation activities are socially motivated, open and collectively participated, complex, and global by nature. The regions need to move toward open innovation, within a human-centered vision of partnerships between public and private sector actors, with universities playing a crucial role.

Regions should be encouraged to develop regional innovation platforms, which act as demand-based service centers and promote the use of international knowledge to implement the Europe 2020 Strategy, smart specialization, and European partnerships according to the interests and needs of regions. For this to happen, we need to apply the new dynamic understanding of regional innovation ecosystems, in which companies, cities, and universities as well as other public and private sector actors (the "Triple Helix") learn to work together in new and creative ways to fully harness their innovative potential.

New innovative practices do not come about by themselves. One major potential is the use of public procurement. The renewing of the European wide rules must increase the strategic agility and activities of municipalities and other public operators as creators of new solutions. Especially the execution of pre-commercial procurement should be reinforced even more in combination with open innovation to speed up the green knowledge society development, i.e., for common reusable solutions in creating the infrastructures and services modern real-world innovation ecosystems are built upon. Conditions must be created that also allow for extensive development projects which address complex societal challenges and which take the form of risk-taking consortia.

One of our working instruments within the Committee of the Regions is the Europe 2020 Monitoring Platform, which broadly reviews and reflects the opinions and decisions on regional level all around Europe. It gives a flavor of cultural and other socioeconomic differences inside the EU. This brings an important perspective to the intellectual capital, namely, the values and attitudes needed for citizens supporting policymakers on appropriate long-term investments and policies.

Emphasizing the importance of these issues, decision makers in all countries and regions worldwide need a deep and broad understanding of the critical success factors affecting the national intellectual capital. With all the facts and frames for thinking, this booklet gives a valuable insight in today's challenges.

Markku Markkula
Advisor to the Aalto University Presidents
Member of the EU Committee of the Regions
Former Member of the Parliament of Finland

Foreword 2

Financial crisis—words very much heard today. What is all this about, actually, and how to get a grip on what we experience today? This booklet gives an important insight on the factors affecting competitiveness and productivity in modern knowledge society. We need to see behind the obvious, and we need to have increasingly "qualified guesses" as the character of the society and industry has fundamentally changed.

What is very important to notice is the shift toward intangible value creation beyond the deterministic phenomena we saw very clearly in the industrial era. Cost drivers were the important ones throughout the industry. Mass production, bigger is better, and very traditional productivity factors were the mantra.

However, the production picture is changing. Increasingly value is created by the intangibles, often services related to the tangible components, and even totally in immaterial value creation, where perceptions and expectations determine the market value of the "extended product." We also see rapid change in organizational forms, we see new type of entrepreneurship growing besides the traditional industry clusters, and we see smart specialization of regions and countries.

This means also that there will be clearly different and complementary roles of the actors in innovation and value creation ecosystems. Large companies, small ones, and even microenterprises together with the public sector are traditionally seen as the active partners in such innovation environments. The real issue in the dynamic markets is however that the end users are increasingly to be taken on board as *active subjects* for innovation and not merely treated as objects, customers. Markets need to be shaped and created in much more dynamic way than ever before. Open innovation beyond cross licensing includes the societal capital as important intangible engine for productivity growth. Innovation happens only when the offering is meeting the demand. Otherwise, we can only speak about inventions or ideas.

We need to have a close look at the intellectual capital and the different factors within it when we design our policy approaches. Short-term investments in process capital (infrastructures) and market capital seem to be very important for the manufacturing base as such, but at the same time, measures for longer-term intellectual capital development and efficiency need to be taken.

Increasingly important is the *structure* and the open *processes* related to intangible capital and knowledge pools. For sustainable long-term development, both the *human capital and renewal capital are crucial*, as they are directly related to the innovation capability of the region. The correlation between these factors and the GDP growth is undisputable. In knowledge intense industries, talent is attracting talent, and the connectivity which modern ICT provides makes this talent pool fluid across disciplines, organizations, and geographical settings. It is imperative to modernize the innovation systems enabling the full dynamics needed for success in knowledge intense industries, beyond the traditional boundaries.

Measuring performance of innovation systems becomes increasingly complex due to the mash-up of different disciplines, having new types of actors and interactions between them. Hence, the importance of analysis of the various components of the national intellectual capital (and equally on national innovation capability) as done in this booklet cannot be underestimated when making qualified guesses for operational choices to create functioning innovation ecosystems. The only predictable in true innovation is the unpredictability and the surprises. The role of the public sector is to drive strategy and measures enabling the unpredictable and to catalyze a fluid, seamless, and frictionless innovation system to grow, with strong interplay with the surrounding society.

We need to have courage to experiment, to prototype in real-world settings, to have all stakeholders involved to find and remove the friction points of innovation, and to achieve sustainable innovation ecosystems for knowledge intense products and services.

I wish you interesting reading with this mind opening report.

Bror Salmelin
Advisor, Innovation Systems
European Commission
DG CONNECT

Foreword 3

The 2008 global financial crisis hit the whole world with unprecedented speed, causing widespread financial panic. Consumer confidence dropped to the lowest level since the Great Depression. Taiwan, with an export-dependent economy, was seriously impacted by the crisis and the unemployment rate hiked, while household consumption levels dropped. At the onset of the financial crisis, Professor Lin was the Dean of Student Affairs here at National Chengchi University in Taipei, Taiwan. She was the dean in charge of financial aid and student loans and thus saw firsthand the direct impact the financial crisis had upon our students. The crisis was so devastating that Professor Lin, along with the university, was compelled to launch several new initiatives to raise money and help students weather the difficult times.

I am very glad that she took this painful experience to heart and set herself upon the task of investigating the impact of the crisis, trying to look into the causes and consequences for policy implications, not only for Taiwan but for an array of 48 countries. In particular, she approaches the crisis from the perspective of "national intellectual capital," which is very important in today's knowledge-driven economy.

Taiwan is an example of a knowledge economy and has enjoyed the fame of being referred to as a "high-tech island." Without an abundance of natural resources, Taiwan's hardworking and highly educated population is the single most precious resource that the island has. Acknowledging the value of such human resources and intellectual capital, we established the Taiwan Intellectual Capital Research Center (TICRC) under my leadership in 2003. Ever since then, Taiwan's government has continuously funded the university to conduct relevant research projects aimed at enhancing the intellectual capital of Taiwan. Having been thus endowed with the responsibility of nourishing future leaders in the public and private sectors, we have focused on building up our strength in innovation, entrepreneurship, and technology management-related research and education.

To enhance intellectual capital research, we recently formed a joint team of professors for a 4-year project in order to leverage their respective research capabilities. Through this project, we hope to provide policy suggestions for the government by exploring the creativity, innovation, and intellectual capital at national, regional, city, and county levels. The goal is to come up with an intangible assets agenda for Taiwan's future sustainability. Professor Lin is an integral member in this research team.

Following her 2011 book *National Intellectual Capital: A Comparison of 40 Countries*, this booklet series is Professor Lin's second attempt at presenting her research, conducted under the sponsorship of TICRC, to international readers. As the Founding Director of TICRC and her President, I am honored to give a brief introduction of the value of this booklet series.

In comparison to her 2011 book, this series increased the number of countries studied to 48 and particularly focuses on the impact of intellectual capital on the 2008 global financial crisis. Rarely has an economic issue been systematically studied from the view point of intangible assets, particularly at such a large scale of 48 countries. The research results show without a doubt that national intellectual capital is indeed an important economic development enhancer. In particular, the fact that countries with higher national intellectual capital experienced faster recoveries from the 2008 financial crisis provides a strong message for the policymakers.

In addition to providing insights to national policy, this booklet also summarizes the background of each country before the crisis, the key events during the crisis, economic development afterward, and future prospects and challenges. Each volume affords readers a holistic picture of what happened in each country in an efficient manner. The linkage between national intellectual capital and this financial crisis also provides a different perspective of the crisis.

We are happy that Professor Lin continues to share her valuable research results with international readers. I sincerely hope that her insights can garner more attention concerning the benefits of developing national intellectual capital for the well-being of every nation.

Se-Hwa Wu
Professor, Graduate Institute of Technology and Innovation Management
President, National Chengchi University, Taipei, Taiwan

Preface 1

There are "mounting risks of a breakup of the Euro zone." Such comments are frequent today on how the European leaders are handling the escalating crisis and its potential impact on non-European countries. But few leaders, reporters, or researchers are actually addressing the situation of national intellectual capital (NIC) and its signals. In addition to the financial crisis, is there an emerging NIC crisis as well? Why is it emerging? How should policymakers think about NIC? In what way does it need specific attention? When will the outcome and impact of taken NIC policy steps be realized?

In the midst of the European crisis, there are national interventions to address the issues mentioned above. In leading economical nations, the investment going into intangibles now exceeds tangibles and is positively correlated to income per capita. However, these still do not show up clearly in national mapping as well as policy-making insights. Therefore, the New Club of Paris is focusing the knowledge agenda setting for countries on societal innovation (see www.new-club-of-paris.org).

Chairman Ben Bernanke of the US Federal Reserve was addressing some of these same aspects in a key note speech in May 2011 hosted by Georgetown University: http://www.icapitaladvisors.com/2011/05/31/bernanke-on-intangible-capital/. OECD and the World Bank are developing NIC statistics, often based on the model from Corrado-Hultén. Japan has been developing both NIC and intangible assets (IA) at METI for some time now. Their research on IC/IA has resulted in a National IA Week with various key stakeholders, such as government agencies, universities, stock exchange, and enterprises. Japan is so far the only country in the world to hold such activities, and they have been doing so for the last 8 years. Australia, Singapore, South Korea, and China are currently undertaking various NIC initiatives. Other countries are also becoming more and more aware of NIC, with policy rhetoric centered on innovation, education, R&D, and trade. Despite this, the map for a more justified NIC navigation has been missing.

This booklet highlights NIC development for a number of countries, based on 28 different indicators, aggregated into four major NIC components of human capital, market capital, process capital, and renewal capital. The model here is a refined and verified statistical model in comparison to the Corrado-Hultén model. We call it the

L-E-S model after the contributors Lin-Edvinsson-Stahle. Based on a deeper understanding and the timeline pattern it sets forth, this model will add to a better NIC navigation, not to mention knowledge agenda setting for countries.

Upon looking at a global cluster NIC map, it is evident that the top leading countries seem to be small countries, especially Singapore, the Nordic countries, Hong Kong, and Taiwan. For the USA, Finland, and Sweden, around 50% or more of its economical growth is related to NIC aspects. Sweden, Finland, Switzerland, the USA, Israel, and Denmark are strongly influenced in its GDP growth by focusing on renewal capital.

It might be that we will see a clearer map of the NIC ecosystem and drivers for wealth emerge in the extension of this ongoing unique research of NIC. This booklet will present a NIC map for various clusters of countries. It can be used for bench marking as well as bench learning for policy prototyping. The starting point is awareness and thinking of NIC and its drivers for economic results. Based on this more refined navigation, NIC metrics can be presented.

Deeper understanding will emerge from this research, such as the scaling up of limited skilled human capital in one nation by using the globalized broadband technologies for migration and flow of knowledge (such as telemedicine or mobile banking in Africa). This is also referred to as the IC multiplier. It might also be the way the old British Commonwealth was constructed but without the IC taxonomy. In modern taxonomy, it might be the shaping of NIC alliances for the migration and flow of IC between nations.

Another understanding that might emerge for policymaking is the issue of employment versus unemployment. The critical understanding will be deployment of IC drivers. This will require another networked workforce of value networkers on a global scale, such as volunteering software and apps developers. However, such volunteers do not show up in traditional statistics, for the mapping on behalf of policymakers.

On another level, there might be a clear gap analyses between nations to support the vision process of a nation. On a deeper level, it is also a leadership responsibility to address the gap of NIC positions versus potential positions. Such a gap is in fact a liability to the citizens, to be addressed in due time.

This will take us to the need for the continuous renewal of social systems. The so-called Arab Spring is explained by some as resulting from three drivers: lack of renewal of social systems, Internet, and soccer as cross-class interaction space. The lack of social renewal and innovation is most likely critical early warning signals. For Greece, we can see such a tipping point occurred back in 1999.

On a global scale, we might see that the concern for the Euro zone crisis should and can be explained by a deeper and supplementary understanding of national intellectual capital, in addition to financial capital. So we need to refine our NIC understanding, NIC mapping, NIC metrics, and NIC organizational constructs into societal innovation for the benefit of wealth creation of subsequent generations.

Leif Edvinsson
The World's First Professor of Intellectual Capital
Chairman and Co-founder of New Club of Paris

Preface 2

Our first book *National Intellectual Capital: A Comparison of 40 Countries* was published in early 2011, at a time when the 2008 global financial crisis had been declared over yet the European region was still plagued with sovereign debt problems. Before we finalized the book, we were able to retrieve some of our raw data concerning the troubled countries, such as Greece, Iceland, Ireland, Portugal, and Spain. The results of our analysis based on data spanning 1995–2008 revealed some early warning signs of the financial turmoil in those countries. In my preface of that book, I mentioned the warning signs might reveal only the tip of an iceberg. At that time, my coauthor, Professor Edvinsson, and I decided to do a follow-up study to trace the development of national intellectual capital (NIC) in as many countries as possible, particularly through the lens of the 2008 global financial crisis. This 12 booklet series is the result of that determination.

The 2008 global financial crisis came with unexpected speed and had such a widespread effect that surprised many countries far from the epicenter of the initial US subprime financial problem, geographically and financially. According to reports, no country was immune from the impact of this financial crisis. Such development clearly signifies how closely connected the world has become and the importance of having a global interdependent view. By reporting what happened during 2005–2010 in 48 major countries throughout the world, this booklet series serves the purpose of uncovering national problems before the crisis, government coping strategies, stimulus plans, potential prospects and challenges of each individual country, and the interdependence between countries. The 6 years of data allow us to compare NIC and economic development crossing before, during, and after the financial crisis. They are handy booklets for readers to have a quick yet overall view of countries of personal interest. The list of 48 countries in 11 clusters is provided in the appendix of each booklet.

Searching for financial crisis-related literature for 48 countries is itself a very daunting task, not to mention summarizing and analyzing it. For financial crisis-related literature, we mainly relied on the reports and statistics of certain world organizations, including OECD, World Bank, United Nations, International Monetary Fund (IMF), European Commission Office, the US Congressional

Research Service, the U.S. Central Intelligence Agency, and International Labor Office (ILO). Some reliable research centers, such as the National Bureau of Economic Research in the U.S., World Economic Forum, the Heritage Foundation in the U.S., and government websites from each country, were also our sources of information. Due to the requirement of more update and comprehensive information, we were not able to use as much academic literature as we would have liked, because it generally covers a very specific topic with time lag and with research methods not easily comprehended by the general public. Therefore, we had to resort to some online news reports for more current information.

In the middle of 2012, the lasting financial troubles caused the European economy to tilt back into a recession, which also slowed down economic growth across the globe. However, almost 4 years have passed since the outbreak of the global financial crisis in late 2008; it is about time to reflect on what happened and the impact of the financial crisis. By comparing so many countries, we came to a preliminary conclusion that countries with faster recovery from the financial crisis have higher national intellectual capital than those with slower recovery. In other words, countries that rebounded fast from the crisis generally have solid NIC fundamentals, including human capital, market capital, process capital, and renewal capital. We also found that the higher the NIC, the higher the GDP per capita (ppp). This booklet series provides a different perspective to look beyond the traditional economic indicators for national development.

In an era when intangible assets have become a key competitive advantage, investing in national intellectual capital development is investing in future national development and well-being.

Enjoy!

Carol Yeh-Yun Lin
Professor, Dept. of Business Administration
National Chengchi University, Taiwan
Taiwan Intellectual Capital Research Center (TICRC)

Executive Summary

The four Greater China economies can capitalize on their synergy to form a value chain and provide a complete spectrum of services to the rest of the world.

Did national intellectual capital (NIC) facilitate China's fast development over the last decade? What are the secrets to the fast recovery of the four Greater China economies from the 2008 global financial crisis? How can Singapore achieve both outstanding NIC and GDP growth with such a tiny area and limited resources?

One of the key factors of the financial crisis was that conventional financial systems failed to detect potential risks due to nontransparent information disclosure. Our earlier NIC research revealed certain warning signs of impending financial crisis for Greece, Iceland, and Ireland. Such findings indicate that NIC, albeit intangible, can provide valuable insights into risk control and strategy formulation. This booklet looks at the connections between the financial crisis and NIC development for China, Hong Kong, Singapore, and Taiwan.

Data covering 2005–2010 for 48 countries indicate that the higher the NIC, the higher the GDP per capital (ppp), accentuating the value of NIC in major countries throughout the world. **For the 6-year average of NIC rankings, China ranks #40, Hong Kong #14, Singapore #5, and Taiwan #18.**

The 2008 financial crisis caused severe impacts all across the globe and is considered to be the worst since the Great Depression of the 1930s. The crisis came with unexpected speed and spread into a global economic shock, which resulted in a number of bank failures. During this period, economies worldwide slowed, credits tightened, and international trade declined. In an effort to mitigate the crisis, governments and central banks worldwide responded with unprecedented fiscal stimuli, monetary policy expansions, and institutional bailouts. These measures had its desired impact, and the financial crisis was declared over by the end of 2009. However, as of the middle of 2012, economic recovery in most developed countries is still hampered by economic slowdown and sovereign debt troubles. In stark contrast with this, the four Greater China economies that this booklet focuses on not only swiftly recovered from the crisis but also steadily progressed over the past few years.

The Greater China economies depend heavily on exports and were hard hit by this global financial crisis, except China. Over the 6 years between 2005 and 2010, Hong Kong, Singapore, and Taiwan had drastic declines in real GDP growth from 2007 to 2009; yet growth for all three economies rebounded quickly in 2010, which is indicative of fast recoveries. China still managed to achieve 9.08% and 8.56% real GDP growth in 2008 and 2009, respectively.

The levels of general government debt in the Greater China economies were relatively low when compared with that of other major nations. In particular, China and Hong Kong did not have much fluctuation in government debt over the 6 years. This means that the financial crisis and aggressive stimulus plans implemented did not significantly impact their government debt levels. Hong Kong actually managed to reduce its debt year by year to less than one percent of GDP by 2010. For the other two economies, both Singapore and Taiwan had to take on additional debt starting in 2008 due to the financial crisis, but their debt level remained under 50% of GDP (lower than the suggested 60% limit for EU countries).

Aside from government debt, unemployment was also a serious problem for the Greater China economies during the financial crisis. At the height of the crisis, Taiwan had the highest unemployment rate, which reached to almost 6%. While this number aroused great concern compared with the country's standard, it is still low when compared with the levels that were reached in certain southern European nations. Singapore started out with the lowest unemployment rate in 2005 (3.1%) and remains the lowest through all 6 years with only 2.2% for 2010. Even in its worst year of 2009, it maintained a small 3% unemployment rate.

In the Greater China economies, Singapore has the highest NIC with market capital, process capital, and financial capital ranking 1, 2, and 2, respectively, among the 48 countries. Hong Kong ranks 2 and 4 in market capital and financial capital, respectively. Such rankings explain that these two economies are both strong in real economy (financial capital) and in intangible assets among the 48 world major countries. Compared to its overall NIC ranking of 40, China stands out particularly in the area of market capital with a ranking of 28. Taiwan has the best renewal capital with a ranking of 10, which bids well for long-term development potential. In general, the NIC levels of the three small democratic economies (except China) are within the top 30% of the 48 countries.

The NIC 3D trajectory analysis was conducted to detect the enhancing and impeding factors of each economy in reaching a targeted GDP per capita (ppp), benchmarking Singapore due to its high financial capital, market capital, and process capital. To reach Singapore's GDP level, for sure China has the longest distance to cover with its being an emerging country. Surprisingly, Hong Kong and Taiwan have very similar distances: -33.62% and –33.52% (see Table 3.4), respectively, to the target GDP, even though Hong Kong has stronger market capital, process capital, and financial capital than those of Taiwan.

As of the mid-2012, the world economy has been slowed down by the pending debt problems in the Euro zone. The economies of Hong Kong, Singapore, and Taiwan were affected and exhibited slower growth. Each economy's resilience to crisis will again be tested if the world plunges back into recession.

This economic crisis provides an ideal opportunity for nations to examine the soundness of their economic system and the effectiveness of national governance related to NIC. The following implications are drawn from our research findings. Readers can refer to Chapter 5 for the rationale behind these implications.

1. The Greater China economies are on solid ground progressing toward continued economic recovery and future development.
2. The Greater China economies have become a powerful integrated economic entity that can sustain itself.
3. The Greater China economies can collaborate in NIC development to reap a synergic effect for their next stage of economic development.
4. Singapore, Hong Kong, and Taiwan can benchmark the Nordic region for long-term human capital and renewal capital development.
5. NIC international competitiveness can be regularly monitored for strategizing national development and well-being.
6. NIC mapping and identification of key efficiency drivers provide directions for linking NIC development to the bottom line GDP growth.
7. Leveraging NIC to secure national well-being is a winning strategy for resource-strained economies.
8. The Greater China economies can capitalize on their synergy to form a value chain and provide a complete spectrum of services to the rest of the world.

In the coming decade, the heart of economic activity will move from the West to the East. East Asia has been named as the growth engine of the future global economy. The Greater China economies reported in this volume are the key contributors of East Asian prosperity. Their resilience and interdependence during the 2008 global financial crisis created a new model for how countries can leverage each other's resources in the future. This report uncovers that the strength of their economy is indeed supported by their NIC. In an era when the intangible asset has become a key competitive advantage, investing in national intellectual capital development is in essence investing in future economic development and well-being. National intellectual capital evolution should be nourished both from local culture viewpoint as well as global interconnectivity by social media.

Based on the emerging new insights of values, societal history as well as citizen relationships, a key focus for the future will be on the fusion of national intellectual capital and social service innovation as well as societal innovation, for the enabling of a new societal fabric.

Contents

List of Figures

List of Tables

Appendix

Chapter 1
Introduction

In today's knowledge economy, knowledge and other intangibles play increasingly important roles, fueling a country's growth and creating significant implications for national value. Knowledge assets represent the source of competencies and capabilities deemed essential for national economic growth, human development, and quality of life (Malhotra 2003). Consequently, countries with deep intangible assets fare better in terms of national wealth and well-being than those whose assets are limited to land, tools, and labor (Malhotra 2003; World Bank 1998). National intellectual capital (NIC) mainly consists of human capital, market capital, process capital, renewal capital, and financial capital. Together, these five types of intellectual capital form a valuable intangible asset and are an important source of national competitive advantage.

The biggest event closely related to NIC in the late 2000s was the 2008 global financial crisis. Financial crisis is a cyclical recurrence in rich and poor countries alike over the past couple of centuries. Each time, the pattern is similar: capital crunch, currency crash, high inflation, high unemployment, undulations in housing and equity prices, and government defaults on international and domestic debts (Reinhart and Rogoff 2011). The sources of problems generally include improper rules and policies, ineffective governance, failed surveillance systems, and implementation flaws. These problems are mainly intangible in nature, which explains the value of examining the connection between NIC and 2008 financial crisis.

The 2008 global financial crisis is considered by many economists to be the worst one in the last 80 years. The crisis rapidly developed and spread into a global economic shock. World political leaders, national ministers of finance, and central bank directors coordinated their efforts to reduce fear, but the crisis continued and eventually led to a global currency crisis. During this period, economies worldwide slowed, credits tightened, and international trade declined.

To curb the impact of the financial crisis, governments and central banks worldwide responded with unprecedented fiscal stimuli, monetary policy expansions, and institutional bailouts in their respective countries. The financial rescue worked, and an economic crisis akin to the Great Depression was avoided. By the end of 2009,

C.Y.-Y. Lin et al., *National Intellectual Capital and the Financial Crisis in China, Hong Kong, Singapore, and Taiwan*, SpringerBriefs in Economics 8,
DOI 10.1007/978-1-4614-5984-2_1, © The Author(s) 2013

the crisis was formally declared over (Kehoe 2010). Yet, while the crisis was officially over, economic recovery in most developed countries continued to trudge along beneath prerecession levels as governments continued to struggle with their finances (Norris 2011).

During the early stage of the financial crisis, management scholars criticized traditional accounting system's inability to reveal the intangibles that explain hidden risks and values for proper decision making (Reavis 2009). Intellectual capital (Edvinsson and Malone 1997) advocates the value of intangible human capital, social capital, and the like, and it has gained increasing attention in today's knowledge-driven economy.

It is our deep belief that national intellectual capital, albeit intangible, can provide valuable insights to policy makers regarding future risk control and strategy formulation. Our previous book, *National Intellectual Capital: A Comparison of 40 Countries* (Lin and Edvinsson 2011; www.nic40.org), was born out of this belief and traces the national intellectual capital development of 40 countries over a 14-year period from 1995 to 2008. The data analysis presented in our previous work revealed certain warning signs of impending financial crisis for countries such as Greece, Iceland, and Ireland (Lin and Edvinsson 2011; 327–333). As a follow-up study, this series is an attempt to further understand the connections between the 2008 global financial crisis and NIC development.

The booklet series, in its entirety, will examine the NIC level of 48 countries during the period from 2005 to 2010 in order to glean new understanding about whether there is a NIC development pattern that distinguishes fast recovery countries from slow recovery ones. This analysis is presented through a series of 11 country clusters, with each booklet focusing on one particular cluster. The clusters are determined based on three factors: geographical proximity, geographical size, or the country's phase of economic development. Focusing on one cluster at a time, we probe the areas of concern within a single country and then compare them between multiple countries to see whether the situation before and after the crisis can be explained by the intangible NIC. Our data comes from the well-recognized International Institute for Management Development (IMD) in Switzerland. The IMD has been publishing yearly rankings of World Competitiveness for around two decades. It is our hope that the insight in this booklet series can provide a different perspective of the financial crisis for future policy implications.

Volume two (this volume) will focus on the four Greater China economies: China, Hong Kong, Singapore, and Taiwan, where there is a majority of ethnic Chinese. The term "Greater China Region" or "Greater China Area" is commonly used in demarcating multinational corporation operation in terms of commercial ties, but not sovereignty. This booklet first provides a brief background on these four economies as a whole before going into each individual economic development. Through this process, the authors hope to paint a general picture of the economic condition and provide a basis for our dataset and analysis in future sections.

Next, Chap. 2 will briefly introduce the impact of 2008 financial crisis on the four economies. Chapter 3 discusses the NIC development of these four economies. Chapter 4 describes issues beyond the financial crisis, and Chap. 5 concludes with future perspectives and policy implications.

Economic Background

This volume is the only one in this series that groups the countries analyzed according to ethnicity. This makes sense in the context of the Greater China economies as the history and economy of this region is interconnected. Normally, Greater China refers to China, Hong Kong, Macau, Singapore, and Taiwan. However, Macau does not have an independent economy that reports its own statistics. As a result, this volume reports on four economies only. In terms of interconnectivity, Hong Kong's population consists of about 95% ethnic Chinese (CIA 2012). It used to be part of China until the cession of Hong Kong Island in 1842, Kowloon in 1860, and New Kowloon in 1898 to the United Kingdom. On July 1, 1997, Hong Kong was handed back to China and became the Hong Kong Special Administrative Region (SAR) of China. Singapore's population is made up of approximately 76.8% ethnic Chinese (CIA 2012). In the early 1800s, Singapore became an important trading port in the Southeast Asian region with the opening of market in China under the Qing Dynasty. At the time, there were many Chinese middlemen in Singapore who handled most of the trade between the European and Asian merchants. By 1827, the Chinese ethnicity that traced its roots to the merchants of the Qing Dynasty became the largest ethnic group in Singapore. Taiwan's population is around 98% ethnic Chinese (CIA 2012). It used to be part of China during the Qing Dynasty until 1895 after the first Sino-Japanese War when Taiwan was ceded to Japan. Taiwan was returned to China after the War World II. In 1945, the Kuomintang(KMT), which were defeated in China's civil war, retreated to Taiwan and became the governing party on the island.

The trading partnerships between the four economies have been tight because of language and culture similarities. This was especially true after China's integration with the world economy in the 1990s. Since the economic history of these four economies go back centuries, it is impossible to cover the entire spectrum in our current work. As such, this background discussion will consider events in the relative recent history from 2005 onward that have the most direct impact upon the current economic conditions of each specific economy. In addition, particular attention will be given to the 2008 global financial crisis. In doing so, the authors hope that the background, in conjunction with our later data and analysis, will provide a "before, during, and after" picture of what was happening from a macroeconomic and intangible assets viewpoint.

After the financial crisis, the four Greater China economies were widely recognized for their fast recovery. To paint a general picture of their global competitiveness in the most recent years and before the financial crisis, we introduce hereunder the Global Competitiveness Index (GCI) published by the World Economic Forum for the reader's reference. This index is relatively robust as it takes into account the 12 distinct pillars[1] containing basic requirements, efficiency enhancers, and

[1]The 12 pillars include institutions, infrastructure, macroeconomic environment, health and primary education, higher education and training, goods market efficiency, labor market efficiency, financial market development, technological readiness, market size, business sophistication, and innovation.

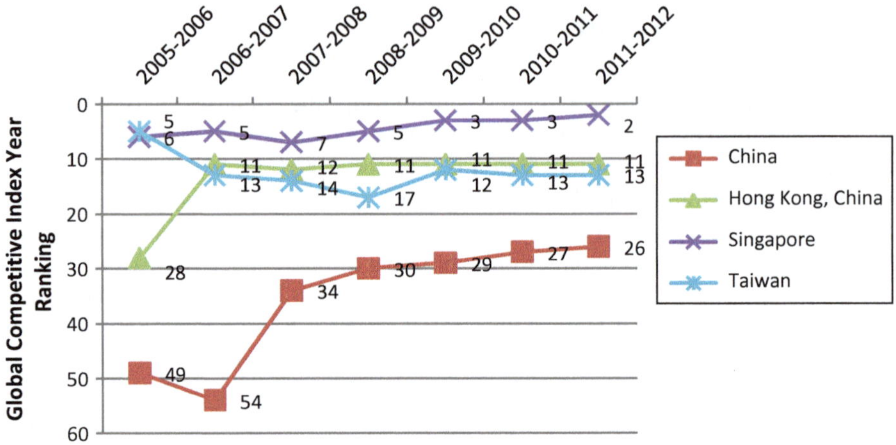

Fig. 1.1 GCI rankings of the four Greater China economies

innovation factors that contribute to a nation's overall economic strength. Based upon commonly accepted economic theory, the development of a total 142 countries was split into three stages in which different factors play the dominant role in determining the outcome of a country's economy. Stage 1, Stage 2, and Stage 3 are respectively characterized by being factor, efficiency, and innovation driven (Schwab 2011). China was categorized a Stage 2 economy, while the other three economies are all Stage 3 economies.

Plotting each economy's annual ranking against a time series of seven periods, Fig. 1.1 displays a rough pictorial overview of the four economies' global competitiveness before and after the financial crisis. As an indicator of economic strength in 2011–2012, it can be seen that Singapore #2, Hong Kong #11, and Taiwan #13 rank within the top 15 worldwide. China, being a Stage 2 economy ranked #26. Despite China's categorization as a Stage 2 economy, its ranking is higher in global competitiveness than even some Stage 3 countries, such as Greece #90, Italy #43, Portugal #45, and Spain #36.

From Fig. 1.1, it can be seen that over the past seven years, the four Greater China economies have each advanced its global competitiveness ranking except for Taiwan. From the last five periods rankings (2007–2008 to 2011–2012), China and Singapore gradually progressed, whereas Hong Kong and Taiwan remained relatively stable. These rankings indicate that the 2008 financial crisis did not have a significant impact on the global competitiveness ranking of each economy with the exception of Taiwan, which experienced slight ranking decline in 2008–2009, yet it bounced back quickly in 2009–2010.

In September 2011, OECD reported that major economies were tilting back into recession as sovereign debt problems continued to ail the Euro zone (Bryant 2011). However, according to the World Economic Forum (Schwab 2011), the growth rate of East Asia is expected to exceed 5% in 2012 despite ongoing economic slowdown in the Euro zone and the United States. For the most recent development, China's gross domestic product (GDP) is estimated to increase by 9.2% in 2012. In March 2012, its large industrial enterprises posted profits up by 25.4% from the previous year and national revenue increased by 24.8%. It is predicted that by 2035, Chinese economy is likely to overcome that of the United States of America (US News 2008).

In contrast to China, the three smaller economies - Singapore, Hong Kong, and Taiwan are all heavily dependent upon exports. With the slowing of the world economy and the lessening of demand for world trade, growth prospects of these three economies are likely to be hampered. Hong Kong's economy is expected to grow 4% in 2012, sliding from 5.75% in 2011 (China Economic Net 2011); Singapore's economy is expected to expand 2.5% in 2012, slowing from 4.9% in 2011 (Reuters 2012); and Taiwan's exports are anticipated to keep growing mildly with an estimated 4.58% growth in 2012 (National Statistics 2012).

The next chapter will give a brief background and qualitative analysis of the financial crisis as it relates to these four economies.

Chapter 2
Impact of the 2008 Global Financial Crisis

In order to present the impact of the 2008 global financial crisis, this chapter will first graphically compare the real GDP growth, total general government debt, unemployment rate, and consumer price inflation of the four economies during the time period from 2005 to 2010. Then, it elaborates on its impact on each economy individually in the sequence of China, Hong Kong, Singapore, and Taiwan.

At the onset of the financial crisis, many countries viewed it as a purely American subprime mortgage problem. Yet, the crisis rapidly developed and spread into a global economic shock. This resulted in the US government bailing out several large financial institutions, such as AIG, Fannie Mae, and Freddie Mac. In addition, coupled with the trouble in America, a number of European banks also failed and stock markets declined across the board (Altman 2009; Fackler 2008). Many Asian countries, far from the epicenter of the financial troubles, felt relatively safe at the beginning. This was especially true for the Greater China economies which held large foreign exchange surpluses at the time. Yet, that view quickly changed as the global economic activities declined over a short period of time. The two largest import regions, the USA and Europe, were in deep financial troubles. This caused international trade to drastically drop, credit to tighten, and direct foreign investments to be swiftly withdrawn, thus resulting in the domino effect of global recession. In the increasingly interconnected world, the Greater China economies were unable to escape unscathed from this financial crisis with their heavy reliance on international trade.

Following the outbreak of the financial crisis, in late 2008, the European Union proposed a European stimulus plan amounting to around US$256 billion or 1.5% of the European Union's GDP—around 1.2% of the GDP from national budgets and 0.3% of the GDP from the EU and European Investment Bank budgets (Europa 2008). For the entire world, the estimated US$2 trillion total in stimulus packages amounted to approximately 3% of the world gross domestic product. This exceeded the call by the International Monetary Fund (IMF) for fiscal stimulus by 2% of the global GDP (Nanto 2009).

C.Y.-Y. Lin et al., *National Intellectual Capital and the Financial Crisis in China, Hong Kong, Singapore, and Taiwan*, SpringerBriefs in Economics 8,
DOI 10.1007/978-1-4614-5984-2_2, © The Author(s) 2013

The impact of the 2008 global financial crisis on each economy can be easily observed from the following four graphs, namely, the percentage of real GDP growth per capita, total general government debt percentage of GDP, unemployment rate of labor force, and consumer price inflation.

Comparisons of the Four Economies

This section presents four graphs in order to examine the Greater China economies as a whole from 2005 to 2010. Figure 2.1 shows that Hong Kong, Singapore, and Taiwan together had a drastic decline in real GDP growth from 2007 to 2009, yet their growth all rebounded quickly in 2010 indicating good recoveries. In the case of China, its percentage of GDP growth per capita fluctuated the least over the six years and managed to still achieve 9.08% and 8.56% growth in 2008 and 2009, respectively. Although Hong Kong had good GDP growth of 5.84% in 2010, it is still lower than its 2005 growth level. Singapore had the earliest decline to negative growth in 2008 and remained in negative growth in 2009 due to its heavy reliance on international trade. However, its rebound in 2010 is the strongest among the four economies, reaching 12.46% and exceeding its 2005 growth level. Taiwan has a fast GDP growth rebound in 2010, reaching 10.62% and exceeding its 2005 growth level as well.

In terms of the total general government debt as a percentage of the GDP, Fig. 2.2 indicates that Singapore has the highest amount of government debt, followed by Taiwan, China, and Hong Kong. The government debt of China and Hong Kong did not have much fluctuation over the six-year period, which means the financial crisis

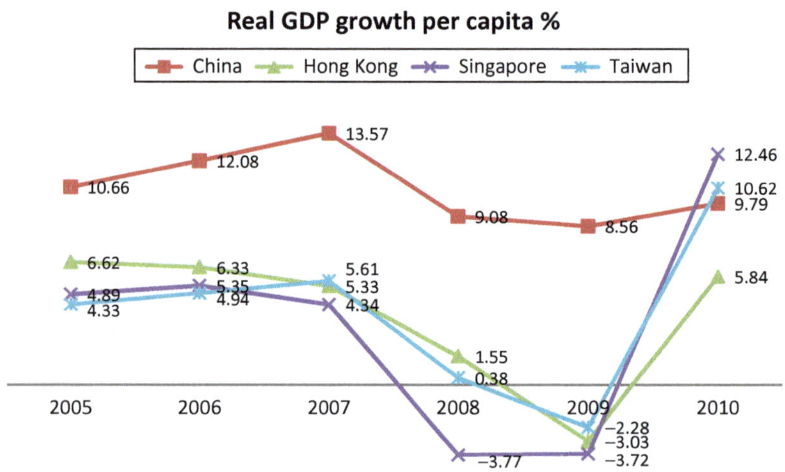

Fig. 2.1 Real GDP growth per capita of China, Hong Kong, Singapore, and Taiwan, 2005–2010

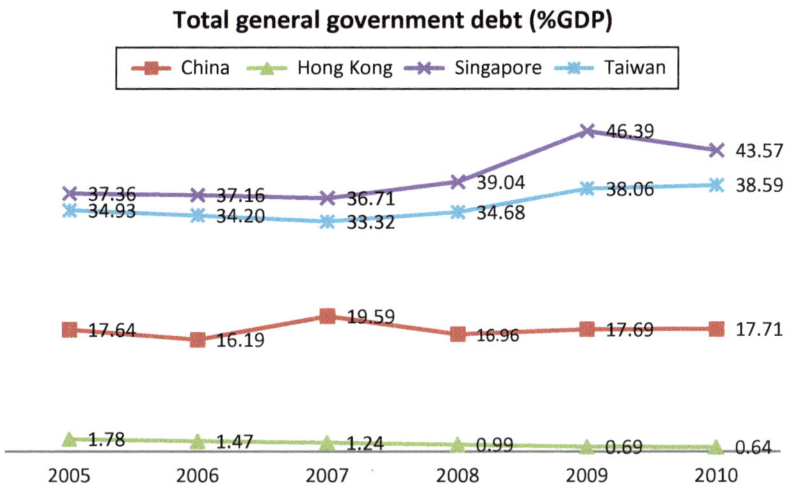

Fig. 2.2 Total general government debt (percentage of GDP) of China, Hong Kong, Singapore, and Taiwan, 2005–2010

did not impact the government debt level. Hong Kong actually reduced its debt year by year, indicating its solid financial status. Singapore and Taiwan had government debt increase starting in 2008, reflecting their increasing financial needs during and after the financial crisis.

Reinhart and Rogoff (2009) reported findings from their research on financial crises over the last 800 years that the aftermath of a financial crisis brings slow and halted growth, sustained high unemployment, and surging public debt—with the overhang of public and private debt being the most important impediment to a normal recovery from the recession.

Figure 2.3 shows the unemployment rate of the four economies. China's unemployment rate is relatively stable, even during the financial crisis. Hong Kong has its lowest unemployment rate (3.6%) in 2008, a sharp increase in 2009 (5.4%), and then a reduction to 4.4% in 2010, which is lower than its 2005 level. Singapore started out with the lowest unemployment rate in 2005 (3.1%) and remained the lowest out of all the economies through the six years. Its large amount of foreign labor (from one out of four to one out of three in 2010) may have served as the buffer and keep its unemployment rate low during the bad times. Taiwan has a clear rising unemployment rate in 2009. Although reduced in 2010, the rate is still higher than its 2005 level.

Figure 2.4 clearly shows the impact of the financial crisis on the general public, as consumer price inflation hiked in 2008 for all four economies, with Singapore being the highest and Taiwan the lowest. This figure also explains the influence of financial stimulus plan as inflation was reduced sharply in 2009, albeit with a slight increase again in 2010. Overall, consumer price inflation in 2010 of these

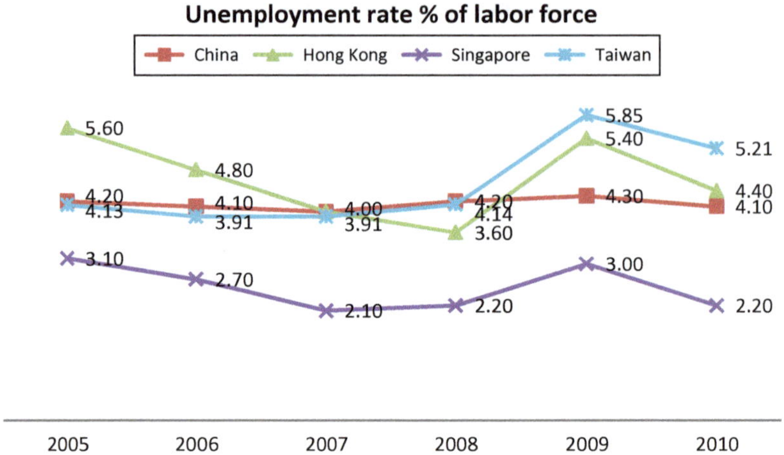

Fig. 2.3 Unemployment rate percentage of labor force for China, Hong Kong, Singapore, and Taiwan, 2005–2010

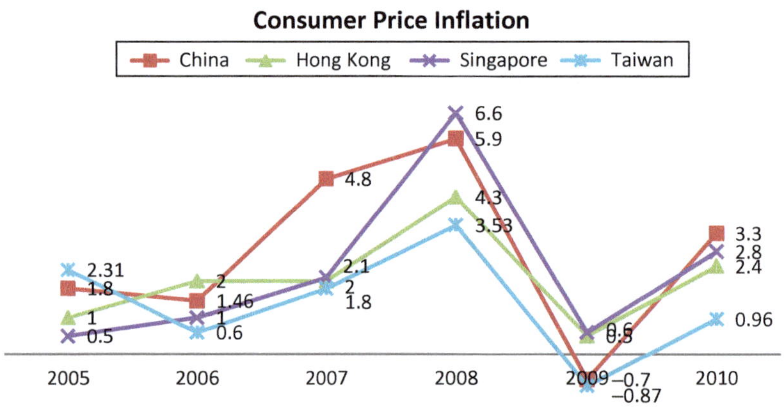

Fig. 2.4 Consumer price inflation of China, Hong Kong, Singapore, and Taiwan, 2005–2010

four economies had not yet returned to their 2005 levels. However, the inflation dropped to about half of 2008 levels explaining consumer markets have been somewhat stabilized.

In general, the four figures shown above indicate that China is relatively unaffected by the financial crisis statistically. The three democratic economies—Hong Kong, Singapore, and Taiwan—were significantly impacted by the crisis, yet they were all able to rebound rapidly.

In what follows, we briefly describe the impact of the 2008 global financial crisis on the four Greater China economies. The depth of the report depends on the English literature available for each economy. For readers to gain a general picture about the efforts that each economy has put in to mitigating the negative impact of the financial crisis, we have summarized the details of stimulus packages implemented by the four economies in Appendix 1. Please note that the reported package is based on publicly available data and is not an exhaustive list. In addition, the reported amounts of stimulus packages were based on the exchange rate at the time of each stimulus, and thus vary. Readers can also refer to Appendix 2 for the important meetings conducted by key global leaders during this financial crisis.

China

China is the fourth largest country in the world in terms of the geographical size after Russia, Canada, and the U.S. Currently, it is also the world's second-largest economy after the U.S. if adjusted for differences in cost of living (CIA 2012). Starting in the 1990s, China's open door policy prompted drastic economic changes that allowed China to be an economic superpower by 2008. A milestone came in December 2001, when China became a member of the World Trade Organization (WTO) (WTO 2012), which accelerated its integration with the world economy. Since then, underpinned by rapid expansions of trade and deep structural changes, its economy has continued to deliver yearly double-digit growth. The speed of its integration with the world economy, coupled with excess demand worldwide, pushed up China's current account surplus to as much as 11% of its GDP by 2007. In 2007, with a trade dependency rate of 66%, China was categorized as having a very open economy by international standards (Heilmann and Schmidt 2010). This high dependency explains why China was not immune from the 2008 global financial crisis.

After the financial crisis hit, thousands of private companies in China closed down. In the first half of 2009, exports sank 21.8% and imports declined 25.4% (Heilmann and Schmidt 2010). The Chinese government estimated that as of 2007 there were 286,200 approved foreign-invested companies in China. Such firms employed more than 42 million people and accounted for 31.5% of gross industrial output value (Morrison 2011). With the international money market crunch, capital flight from emerging countries is a common practice. The speed and scope of this hit came as a shock for Chinese policy makers. In an effort to mitigate the crisis, Chinese president Hu Jintao pledged an economic stimulus measure worth US$586 billion (RMB 4 trillion) in early November of 2008, which amounted to 15.5% of the country's GDP in 2007 (Heilmann and Schmidt 2010). In addition, China also implemented measures such as increasing investments in public infrastructure, loosening monetary policies to increase bank lending and providing various incentives to boost domestic consumption (Morrison 2011). It is important to note that the size of China's stimulus package is comparable to that of the United States' when its GDP is only a third as large (Fleet 2010). Herd et al. (2011) also reported that China responded to the 2008 crisis with fiscal stimulus far greater than that of other OECD countries.

China launched its Economic Stimulus Plan mainly to expand the public sector, including pumping more public investments into infrastructure for rail network, roads, and port development. It also created measures to increase affordable housing, lower taxes on real estate sales and commodities, and ease credit restrictions for mortgage as well as small and medium enterprises (Fleet 2010). China's stimulus funding differs significantly from that in other countries, the central government contributes only about one-fourth of all funds by issuing bonds over a two-year period from 2008 to 2010. The remaining three quarters was to be provided by local governments, state-controlled enterprises, and the market (i.e., government-linked financial institutions and nonpublic firms under government guidance) (Heilmann and Schmidt 2010). That is, a large part of the stimulus package involved off-budget expenditure by local authorities, which resulted in the public expenditure rising by nearly 3% of GDP in 2009.

Critics pointed out that the share of the package was aimed at major infrastructure projects, as opposed to direct stimulus of consumption, and the RMB 4 trillion stimulus package was more a policy than a package (Fleet 2010). Nonetheless, the package worked and recovery started to show by the second half of 2009. This timeframe preceded that of the other major economies. As a result, companies were hiring again, restaurants were filling up, and consumption statistics were turning upward. After the crisis, both the estimated output gap and the OECD composite leading indicators suggested that China is operating above capacity with its ongoing rapid economic transformation (Herd et al. 2011).

Looking back at the stimulus decision-making process, the experience of focusing on public expenditure in the 1997 Asian financial crisis gave the 2008 Beijing policymakers valuable lessons to draw from. Beijing designed more than 50% of the entire RMB 4 trillion in the initial stimulus package to focus on infrastructure. In this manner, the money could be spent quickly and had a significant impact on employment where it was needed the most since many low-skilled workers had lost low-end export-processing jobs. In reality, much of the money actually found its way to private firms who were subcontractors to the initial state-owned enterprise beneficiaries, and it had a huge impact on consumer confidence (Fleet 2010). In addition, China included in its stimulus package a massive increase in liquidity; i.e., funds from the national government would not exceed around a quarter of the designated RMB 4 trillion. Also, the package stipulated that 40% of the amount would be provided by bank lending matched with reduced reserve requirements and interest rates cut (Heilmann and Schmidt 2010). Furthermore, the measures consisted of substantial regulatory change, including a lowering of down payments required for mortgages, reduction or elimination of value added tax (VAT), lower fuel prices and subsidies for smaller cars (fuel efficiency), and an expansion of subsidies for consumer goods purchases in rural areas.

In a knowledge forum conducted by the Wharton school at the University of Pennsylvania, Chou (Khemka et al. 2008) commented that the financial crisis was not a bad thing for China as its economy was growing too fast and needed to be held in check. The preoccupation of the Chinese government for the last few years before the crisis had been how to cool down the economy and the real estate prices, which had become out of the reach of the common people. As far as China can, it had been trying to move away from being overly reliant on the exports sector of the economy. Therefore, the financial crisis offered an opportunity for China to adjust its economy to be more independent and resilient to future external impact.

In general, the fast and decisive government stimulus measures worked well in China. In addition to its own recovery, China also helped pull the world economy along during this financial crisis (Fleet 2010).

Hong Kong

Currently, Hong Kong is a Special Administrative Region (SAR) of the People's Republic of China. In the July 1997 handover agreement, China promised that its socialist economic system would not be imposed on Hong Kong under its "one country, two systems" formula and that Hong Kong would enjoy a high degree of autonomy in all matters except foreign and defense affairs for the next 50 years (CIA 2012).

Ever since, China has been Hong Kong's largest trading partner, accounting for about half of Hong Kong's exports by value. During the past decade, as Hong Kong's manufacturing industry moved to China, its service industry has grown rapidly and accounted for more than 90% of the territory's GDP in 2009. Hong Kong has a free-market economy highly dependent on international trade and finance. The value of goods and services trade in Hong Kong, including reexports, is about four times its GDP. Heavily dependent on trading, Hong Kong was hit hard by the financial crisis in terms of exports, GDP, and employment. Impacted by the crisis, economic growth dropped from 6.4% in 2007 to 2.5% in 2008 and then to further negative growth of -2.5% in the last quarter 2009, which was Hong Kong's worst performance since 1999 (Yang and Tong 2009). Total exports dropped by 21.8% year on year in January 2009, and the stock market saw a sharp decline with the Hang Seng index plunging 48% from a year before, unseen since 1974 (Yang and Tong 2009).

In response to the crisis, the Hong Kong government swiftly put forward a series of measures to stabilize the financial market, support enterprise, and create employment. Externally, China's central government also announced measures to boost Hong Kong's economy through cross-boundary infrastructure projects. In its 2009/2010 budget, the Hong Kong government proposed an expenditure of around US$38.4 billion (HKD 300 billion) and set a budget deficit of about US$5.1 billion (HKD 40 billion), 2.4% of the GDP, yet it is considered by many as conservative given Hong Kong's ample fiscal resources (Yang and Tong 2009).

With a service economy, Hong Kong's financial sector was hard hit by the crisis as output and employment declined. Fortunately, most banks in Hong Kong are sufficiently capitalized and have raised their provision levels so they were able to weather the crisis (Deutsche Bank Research 2009). In addition, Hong Kong's sound banking system, minimal public debt, strong legal system, ample foreign exchange reserves, rigorous anticorruption measures, and close ties with China, enabled it to quickly respond to the financial crisis.

While Hong Kong's open economy left it exposed to the global economic slowdown, its increasing integration with China has mitigated the effect of the financial crisis. Trade, tourism, and financial links between the two economies helped Hong Kong recover quicker than many outside observers anticipated. Particularly, foreign investors continued to contribute capital inflows as they took advantage of Hong Kong's "free"-market economy with high accessibility to the China mainland economy (Tuan and Ng 2011). Although Hong Kong's GDP fell in 2009 as a result of the global financial crisis, a recovery began in the third quarter 2009, and the economy grew nearly 6.8% in 2010 (CIA 2012).

Singapore

For decades, Singapore had a highly developed and successful free-market economy. It enjoyed a remarkably open and corruption-free environment with stable prices and a per capita GDP higher than that of most developed countries. The economy depends heavily on exports, particularly in consumer electronics, information technology products, pharmaceuticals, and financial services. As a result of all this, Singapore's real GDP growth averaged 7.1% between 2004 and 2007 (CIA 2012).

When the 2008 global economic crisis hit, Singapore was the first East Asian country to fall into recession (Thangavelu 2009). This clearly showed the vulnerability of the trade-dependent Singaporean economy. The manufacturing sector was hit particularly hard due to falling demand induced by the overall deterioration of economic conditions in the U.S. and Europe (key export destinations which account for nearly 33% of the total Singapore's non-oil exports over the last few years) (Thangavelu 2009).

Fortunately, due to its well-regulated market, the exposure of Singapore's banks to subprime mortgage was limited (Thangavelu 2009). Even though damage to the banking sector was limited, Singapore still suffered a huge loss of wealth as the stock market plummeted from a high of 3500 points in December 2007 to 1700 points in the last quarter of 2008. This depressed domestic demand and investment in assets (Kesavapany 2010).

Singapore's policy makers responded with a series of measures aimed at insulating the population at large from the negative effects of the financial crisis. The Monetary Authority of Singapore (MAS) guaranteed bank deposits and flushed the money market with enough liquidity to restore market confidence. It established a US$30 billion swap line with the US Federal Reserve to abate any strain in USD funding in the Asian dollar market (Kesavapany 2010). To assist local, small- and medium-sized companies in obtaining access to credit, the Singaporean government established a US$1.56 billion (S$2.3 billion) loan facility and launched a risk-sharing initiative which took on significant shares of bank-lending risks in November 2008

(Kesavapany 2010; Thangavelu 2009). Moreover, the Singapore central bank shifted its currency policy to a "zero-percent appreciation" stance and devalued the Singapore dollar (S$) to help electronics exports (Kesavapany 2010).

The Singaporean government also set aside US$406.2 million (S$600 million) for training and development of workers to improve its human capital (Thangavelu 2009). This had a strong impact on future reemployment and retention of productive workers in the labor market. Furthermore, the government delivered a US$13.8 billion (S$20.5 billion) resilience package for the 2009 financial year for temporary measures to combat the effects of the financial crisis (Abidin 2010). The key objective was to help Singaporeans keep their jobs and thus increase overall confidence. Measures included strategic industrial support in order to drive down business costs and avoid corporate failures as well as income tax rebates to ensure household flexibility without locking down rates (Kesavapany 2010). In addition, Singapore's government spent heavily on big investment projects in the construction sector such as the construction of two integrated resorts, the Marina Bay financial sector and the new MRT line (Kesavapany 2010).

In many ways, Singapore is well positioned to ride out what is likely to be a prolonged global recession for the following reasons: first, the Singaporean government is swift in initiating effective coping measures. Second, its public finances are sound and can run budget deficits without increasing the tax burden. Third, the large pool of foreign workers, about one in every four workers in Singapore, provides a buffer against a sharp rise in unemployment. Fourth, the flexible wage system allows employers to adjust costs better and reduce job cuts (Fong 2009).

In general, Singapore's economy contracted 1.3% in 2009 as a result of the global financial crisis, yet rebounded nearly 14.7% in 2010, on the strength of renewed exports (CIA 2012).

Taiwan

Since the 1970s, Taiwan has continued to develop as a dynamic capitalistic economy with gradually decreasing government guidance in investments and foreign trade. In keeping with the free-market trend, a number of large state-owned banks and industrial firms have been privatized. Taiwan runs a large trade surplus, and its foreign reserve was the fourth largest in 2011, behind only China, Japan, and Russia. Since 2005, China has overtaken the U.S. to become Taiwan's second-largest source of imports after Japan. China is also Taiwan's number one destination for foreign direct investment (CIA 2012).

With a small internal market of only 23 million, fully 70% of Taiwan's gross domestic product is accounted for by exports and the lion's share of these exports consists of electronics, machinery, and information technology products (CIA 2012; Cooke 2012). Heavy dependence upon exports exposes Taiwan to upturns and downturns in world demand. As a result, during the 2008 financial turmoil, Taiwan's unemployment reached levels not seen since 2003, and the economy fell 8.36% in the fourth quarter of 2008 while its GDP contracted 1.9% in 2009, due primarily to a 20% year-on-year decline in exports (QFinance 2011). In response, Taiwan's government launched a US$5.6 billion Economic Vitalization Package in September 2008 (about 3% of its GDP) (QFinance 2011). The stimulus package focused on infrastructure development, low-income households, tax breaks for new investments, and financial incentives for small- and medium-sized businesses. It also aimed at boosting exports to new overseas markets, such as Russia, Brazil, and the Middle East (QFinance 2011).

The Economic Vitalization Package implemented a series of monetary policies, financial stability measures, and fiscal policies in order to increase domestic demand, stabilize the financial system, and maintain economic growth. To increase domestic demand and provide a sufficient injection of liquidity into the market, Taiwan's central bank adopted an easy monetary stance. Key measures included reducing bank base rates by 2.375% through seven cuts between September 2008 and February 2009, to a record low of 1.25% (QFinance 2011). To stabilize the financial markets, improve confidence, and assist individual and corporate funding, Taiwan's government guaranteed all deposits in insured financial institutions by their full amount until the end of 2009. This measure effectively stabilized the market and restored the confidence of depositors. The impact of global financial turmoil also resulted in acute slumps in the local stock markets. To stabilize domestic equity prices, Taiwan temporarily resumed the ban of short selling 150 listed shares below the previous day's closing price and suspended borrowed stocks from short selling. Such policies helped restore the stock market order and reduced man-made manipulation.

In January 2009, the government followed its initial stimulus with a second round of measures entailing the distribution of around US$108 (NTD3600) in shopping vouchers to each of its residents. In addition, the government also launched a number of public construction projects and provided incentives to encourage further private investments. In total, this second package was estimated to worth approximately US$14.9 billion over the four subsequent years.

Chapter 3
National Intellectual Capital Development of the Four Greater China Economies

In this chapter, we first present the development of national intellectual capital (NIC) in the four Greater China economies over the 6-year timeframe and then graphically display the relationship between each individual component of NIC with GDP per capita (ppp). Afterward, we describe the relative positions of long-term and short-term intellectual capital in these four economies and explain the dynamics of NIC for three time periods (2005–2006, 2007–2008, and 2009–2010). Finally, we illustrate the NIC trajectory of these four economies on three-dimensional (3D) 48-country landscapes.

National Intellectual Capital Development

The NIC measurement model in this study is based upon the indicators described in Appendix 3, containing human capital, market capital, process capital, renewal capital, and financial capital. This model has been statistically proven for its validity and reliability. The definitions of the indicators are listed in Appendix 4. This booklet series covers a total of 48 countries in 11 country clusters (see Appendix 5 for details).

For a full picture of all the 48 countries, Appendix 6 exhibits the scores and rankings of human capital, market capital, process capital, renewal capital, financial capital, and overall NIC covered in this booklet series. Table 3.1 summarizes the status of the four Greater China economies compared with the total 48 countries reported in Appendix 6. Appendix 7 provides the individual economic profiles with additional statistics. Appendix 8 outlines the trade dependency within the four economies. For the NIC ranking, Singapore ranks 5, followed by Hong Kong at 14, Taiwan at 18, and China at 40.

From Table 3.1, Singapore excels in market capital (#1), process capital (#2), and financial capital (#2); Hong Kong performs well in market capital (#2), process capital (#5), and financial capital (#4). Taiwan does better in renewal capital (#10)

C.Y.-Y. Lin et al., *National Intellectual Capital and the Financial Crisis in China, Hong Kong, Singapore, and Taiwan*, SpringerBriefs in Economics 8,
DOI 10.1007/978-1-4614-5984-2_3, © The Author(s) 2013

Table 3.1 National intellectual capital scores and ranking of China, Hong Kong, Singapore, and Taiwan among 48 countries spanning 2005–2010

	Human capital		Market capital		Process capital		Renewal capital		Financial capital		NIC	
Country	Score	Ranking	Score	Ranking	Score	Ranking	Score	Ranking	Score	Ranking	Score	Ranking
China	4.942	39	5.184	28	3.746	40	2.099	33	7.967	44	23.938	40
Hong Kong	6.564	25	8.090	2	7.209	5	3.277	23	9.794	4	34.934	14
Singapore	7.144	17	8.306	1	7.506	2	5.416	11	9.961	2	38.334	5
Taiwan	7.244	13	5.823	17	6.026	20	5.425	10	9.526	20	34.044	18

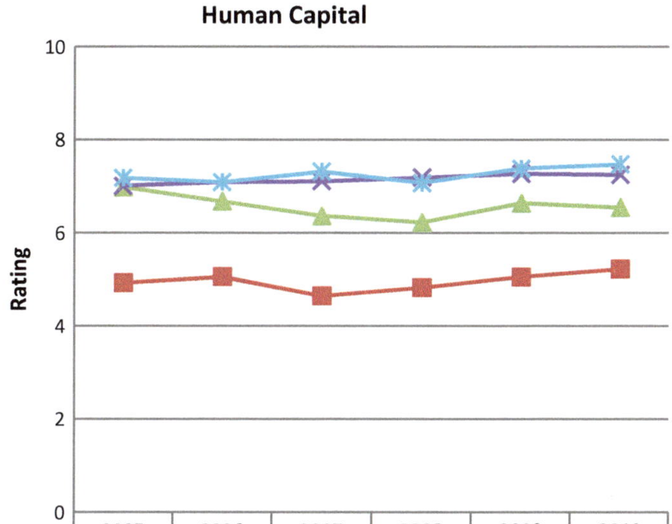

	2005	2006	2007	2008	2009	2010
China	4.92	5.04	4.63	4.81	5.04	5.21
Hong Kong	6.97	6.67	6.36	6.22	6.63	6.54
Singapore	7.00	7.08	7.10	7.18	7.26	7.24
Taiwan	7.17	7.08	7.31	7.06	7.37	7.47

Fig. 3.1 Human capital of China, Hong Kong, Singapore, and Taiwan

and human capital (#13), while China is comparatively strong in market capital (#28) when compared to its overall NIC ranking of 40. Although China is relatively weak in intangible NIC, its affiliate—Hong Kong—is strong in some critical capitals, which should result in synergic effect in the long run.

The following six figures compare human capital, market capital, process capital, renewal capital, financial capital, and NIC for the four economies. All individual capital scores are based on a 1–10 scale with the NIC score being the sum of the five capital scores.

Human Capital

National human capital represents a nation's investment in nurturing qualified human resources, such as higher education enrollment, skilled labor, and public expenditure in education. Human capital is a long-term-oriented NIC (Lin and Edvinsson 2011: 323), which takes time to develop and is not susceptible to external impact. Figure 3.1 shows that human capital of these four economies remained

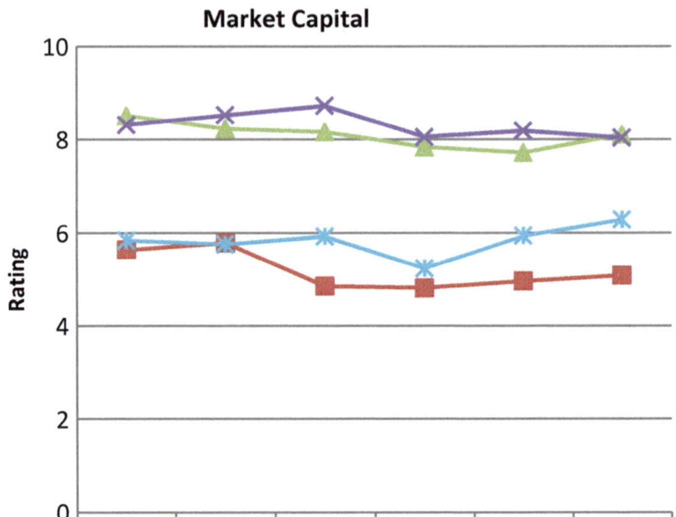

	2005	2006	2007	2008	2009	2010
China	5.63	5.78	4.85	4.81	4.95	5.08
Hong Kong	8.51	8.23	8.16	7.83	7.71	8.10
Singapore	8.31	8.52	8.73	8.05	8.19	8.04
Taiwan	5.83	5.75	5.92	5.23	5.93	6.28

Fig. 3.2 Market capital of China, Hong Kong, Singapore, and Taiwan

relatively stable over the 6 years. Human capital in each of the three democratic economies—Hong Kong, Singapore, and Taiwan—is higher than that of China, with China lagging behind Singapore and Taiwan by more than twenty percent. Hong Kong's human capital in 2005 is similar to that of Singapore and Taiwan, yet it declined slightly over the years.

Market Capital

National market capital represents international trade and relationships such as cross-border ventures, globalization, and exports of goods. Figure 3.2 shows two groups, Hong Kong and Singapore, that are the high-market capital group, whereas China and Taiwan are the low-market capital group. China has an earlier market capital decline in 2007 and remains at a similar level during and after the financial crisis. All three democratic economies have market capital decreased in 2008, demonstrating their heavy reliance on international trade. Yet, they were all able to successfully bounce back after the financial crisis. Among the four, Taiwan has the most obvious market capital rebound, surpassing its 2005 level.

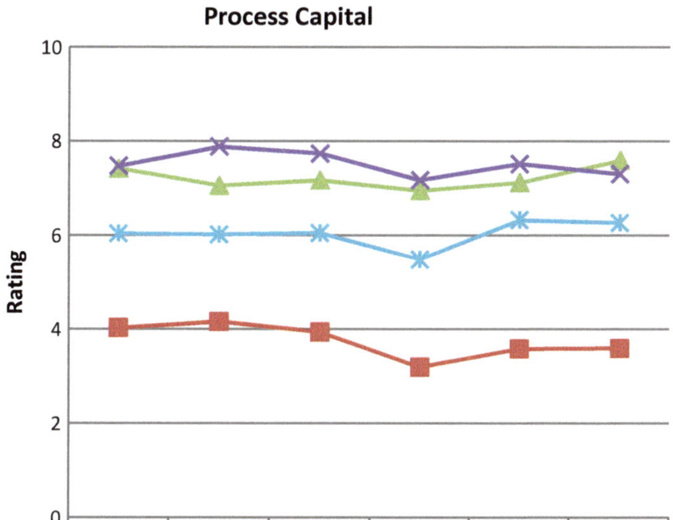

	2005	2006	2007	2008	2009	2010
China	4.03	4.15	3.93	3.19	3.58	3.59
Hong Kong	7.41	7.05	7.16	6.94	7.11	7.58
Singapore	7.46	7.87	7.73	7.16	7.51	7.29
Taiwan	6.03	6.01	6.05	5.48	6.32	6.26

Fig. 3.3 Process capital of China, Hong Kong, Singapore, and Taiwan

Process Capital

National process capital represents the infrastructure required for building a prosperous society, such as capital availability, mobile phone subscribers, and government efficiency. Figure 3.3 indicates that Singapore has the highest process capital, followed by Hong Kong, Taiwan, and China. Yet, Hong Kong surpassed Singapore in 2010. For each of the four economies, there is process capital decline in 2008 followed by an increase in 2009, which shows the recovery taking place after just one year. Hong Kong has the least process capital fluctuation among the four over the 6-year period.

Renewal Capital

National renewal capital represents the capability of R&D and innovation, such as R&D spending, scientific articles, and patents. Figure 3.4 shows that China has the lowest renewal capital level among the four economies. Hong Kong's renewal

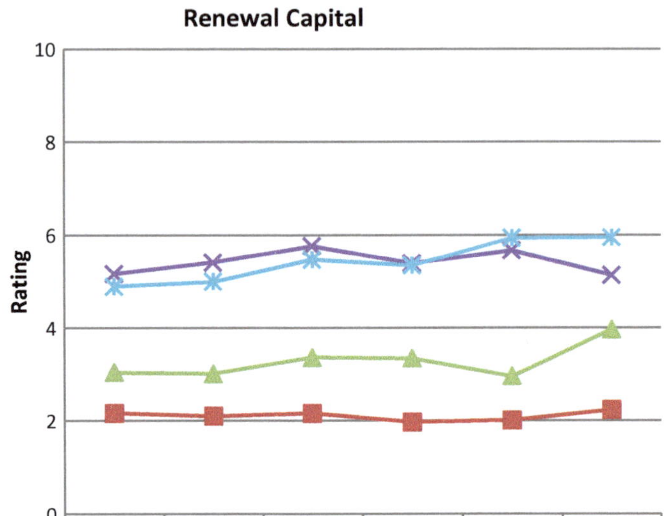

	2005	2006	2007	2008	2009	2010
China	2.15	2.09	2.15	1.97	2.00	2.23
Hong Kong	3.04	3.01	3.36	3.34	2.96	3.96
Singapore	5.16	5.40	5.76	5.38	5.67	5.13
Taiwan	4.88	4.99	5.46	5.33	5.93	5.95

Fig. 3.4 Renewal capital of China, Hong Kong, Singapore, and Taiwan

capital was better than that of China, yet was still lower than Singapore and Taiwan by almost twenty percent before 2009. However, Hong Kong exhibited strong renewal capital improvement in 2010. Singapore and Taiwan were at similar renewal capital levels, yet Taiwan surpassed Singapore starting in 2009. Taiwan had the highest renewal capital in 2010, reflecting its well-recognized innovation, as it placed first in utility patents granted—the fourth straight year it topped the list—according to the World Economic Forum 2011 report (Schwab 2011).

Financial Capital

Financial capital is representative of GDP per capita (ppp) in US dollars transformed to a score on a scale of 1–10. The process of score transformation is explained in Appendix 3. Figure 3.5 shows that China lags far behind the other three democratic economies in financial capital. In real dollars, the average GDP per capita (ppp) of 2005–2010 was US$5,817 for China, US$41,519 for Hong Kong, US$49,743 for Singapore, and US$31,027 for Taiwan.

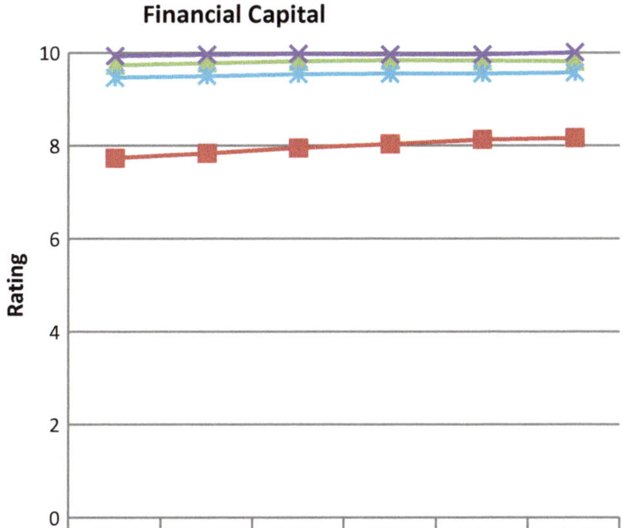

	2005	2006	2007	2008	2009	2010
China	7.73	7.83	7.94	8.03	8.12	8.16
Hong Kong SAR	9.73	9.77	9.81	9.83	9.82	9.81
Singapore	9.93	9.95	9.97	9.96	9.96	10.00
Taiwan	9.46	9.50	9.53	9.54	9.55	9.57

Fig. 3.5 Financial capital of China, Hong Kong, Singapore, and Taiwan

NIC

NIC is the overall national intellectual capital comprised of the five types of capital described in the sections above. Figure 3.6 presents the trends of overall NIC. Singapore has the highest NIC, followed by Hong Kong, Taiwan, and finally China. Except for Hong Kong, the other three economies show NIC declining in 2008, yet bouncing back in 2009. From the relatively flat lines, it can be seen that the NIC for these four economies did not exhibit much fluctuation over the 6 years in spite of the economic downturn. In other words, their valuable intangible assets were not affected much by the financial crisis.

Relationship Between Each Individual Capital and GDP per Capita (ppp)

We investigate the relationship between each individual capital and economic growth by using the following six figures (Figs. 3.7, 3.8, 3.9, 3.10, 3.11, and 3.12). The graphs have four dimensions: the x-axis changes from graph to graph (NIC,

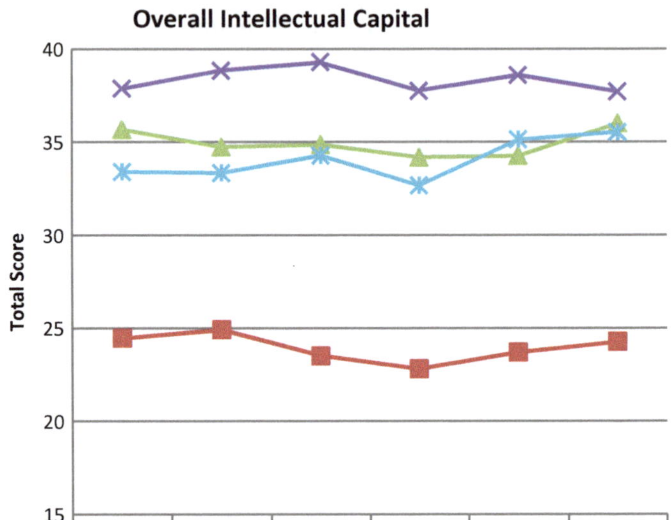

	2005	2006	2007	2008	2009	2010
China	24.45	24.90	23.50	22.81	23.70	24.26
Hong Kong	35.65	34.72	34.85	34.16	34.23	35.99
Singapore	37.86	38.83	39.28	37.74	38.59	37.70
Taiwan	33.38	33.33	34.27	32.66	35.11	35.52

Fig. 3.6 NIC of China, Hong Kong, Singapore, and Taiwan

human capital, market capital, process capital, and renewal capital), the y-axis is a fixed dimension of GDP per capita (ppp) in US dollars (hereafter termed GDP), the color shade represents renewal capital with red denoting higher levels of renewal capital, and the bubble size denotes the population of an economy. Figure 3.7 is the relationship between NIC and GDP for all 48 countries. Figures 3.8, 3.9, 3.10, 3.11, and 3.12 display the codevelopment of the intellectual capital-GDP relationship over the 6-year timeframe for the four economies. Since financial capital is the transformed 1–10 score from GDP, there is no need to plot its relationship against itself. In the figures, 2005 is the starting point tagged with the name of the economy, and the end point is 2010.

In particular, Fig. 3.7 is plotted to clearly show a full picture of the relationship between NIC and GDP for 48 countries in year 2010. The strong positive correlation indicates that the higher the NIC the higher the GDP. Consequently, enhancing NIC is definitely a strategic issue that deserves the attention of national policy makers.

With the data of all 48 countries, Fig. 3.8 shows the formation of the NIC and GDP codevelopment continuum from the bottom left to the upper right. All paths of the four economies fall within the continuum. The path of Singapore is at the upper right corner, Hong Kong and Taiwan in between, and China at the bottom left corner. Apparently, Singapore has higher NIC and higher GDP than the other

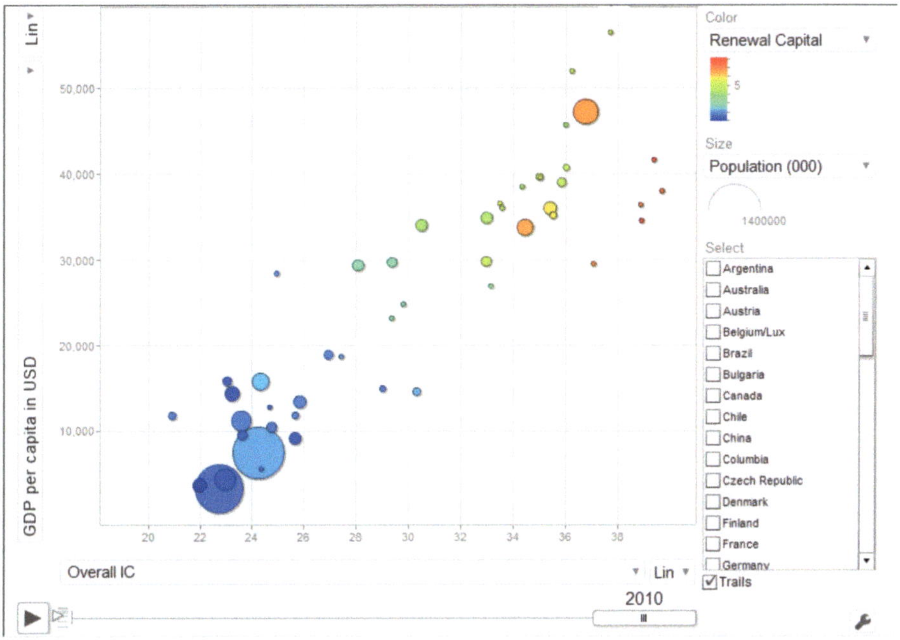

Fig. 3.7 NIC versus GDP per capita (ppp) for 48 countries in 2010

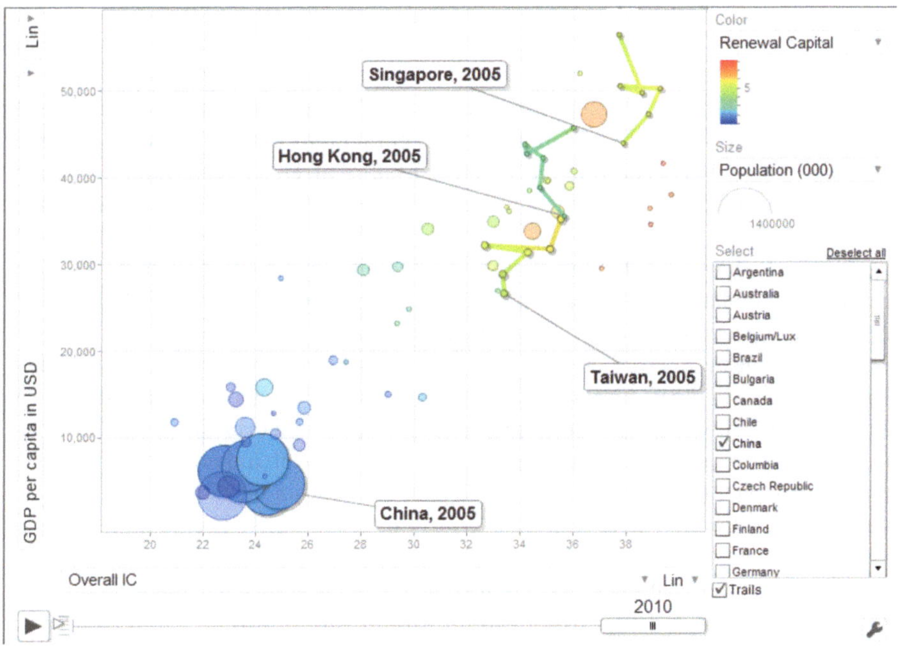

Fig. 3.8 The development of NIC and GDP per capita (ppp) for the Greater China economies, 2005–2010

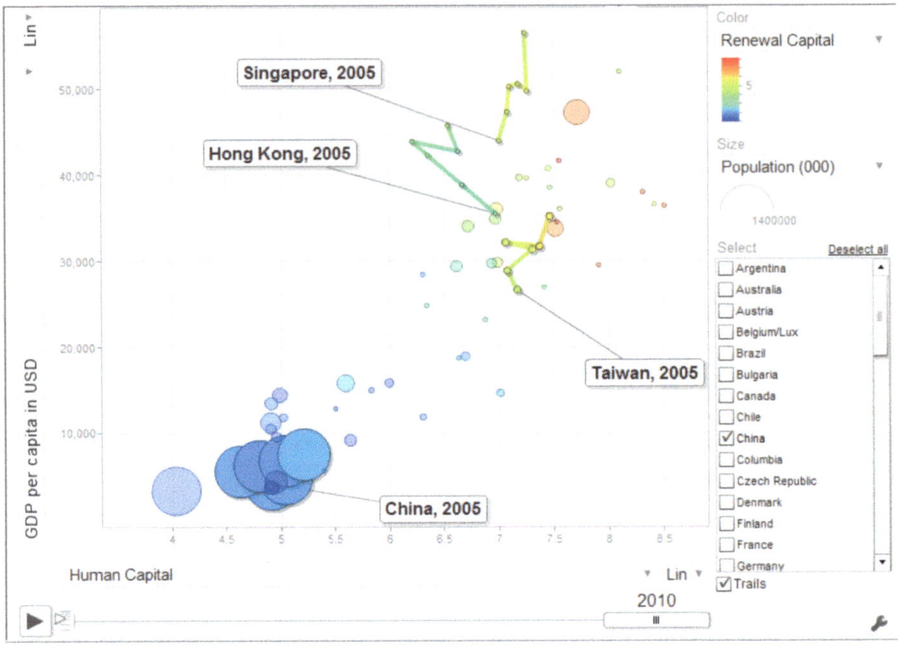

Fig. 3.9 The development of human capital and GDP per capita (ppp) for the Greater China economies, 2005–2010

three Greater China economies. On the other hand, China has lower NIC and lower GDP than not just the other 3 Greater China economies but also most of the other 48 countries as well. The relatively vertical paths of Hong Kong and Singapore indicate that these two economies have faster GDP growth than NIC advancement over the years.

Figure 3.9 shows the "human capital-GDP" continuum formation of the 48 countries and the codevelopment paths of the four economies. The paths of Hong Kong and Singapore are not within the continuum; rather, they are to the left, tilted toward GDP. This can be interpreted as saying that their GDP develops better than their human capital. In 2005, human capital of Hong Kong, Singapore, and Taiwan was at a similar level, yet Hong Kong's human capital declined over the years. However, its GDP kept on rising, meaning Hong Kong's GDP growth did not accompany with its human capital improvement. Taiwan's path falls within the continuum with both human capital improvement and GDP growth over the years. The human capital of China improved over the years, yet its GDP was still low at the bottom.

The "market capital-GDP" formation of Fig. 3.10 shows that there is no clear continuum like the last two figures as the data points spread further apart, which means that the correlation between market capital and GDP is not high for these 48 countries as a whole. The market capital-GDP codevelopment paths of Hong Kong and Singapore are at the high-high, Taiwan at the middle-middle, and China at the

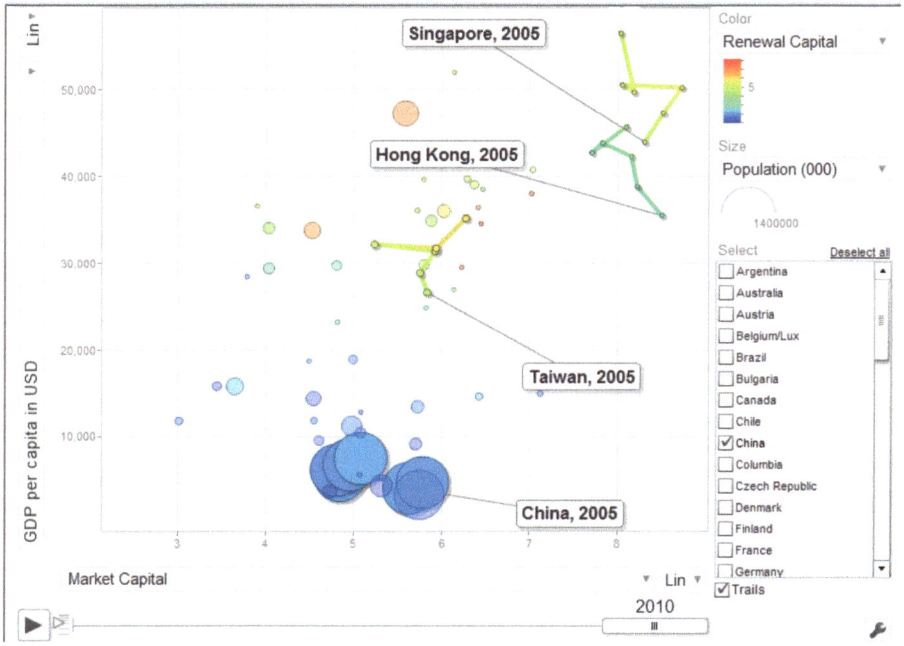

Fig. 3.10 The development of market capital and GDP per capita (ppp) for the Greater China economies, 2005–2010

middle-low part of the graph. For Hong Kong and Singapore, their GDP kept growing even though their market capital declined. For Taiwan, its market capital rebounded after the financial crisis and surpassed its 2005 level. The market capital of China declined during the financial crisis and had not yet recovered to its 2005 level in 2010.

Figure 3.11 shows that the belt-shaped continuum appears again. All the four process capital-GDP codevelopment paths fall within the continuum, with Hong Kong and Singapore at the high-high, Taiwan at the middle-middle, and China at the low-low part of the graph. The process capital improvement goes along with GDP growth for all the four economies, albeit in varying degrees. Both Hong Kong and Taiwan showed clear improvement in process capital after the financial crisis.

Figure 3.12 shows that there is no clear continuum for the 48 countries. The renewal capital-GDP codevelopment path of the four economies is very different from the previous figures. They are spread out in different parts of the graph. If the graph is split into two parts with a dotted line, the paths of Hong Kong and Singapore are tilted toward GDP, which means their GDP grows faster than their renewal capital. Unlike previous figures where the path of Hong Kong followed closely that of Singapore's, it is far behind in this figure. The path of Taiwan is slightly tilted toward renewal capital. Taiwan's path indicates that its renewal capital correlates better with GDP growth than the other three economies. China's path is still far away from that of the others, with little renewal capital and GDP improvement.

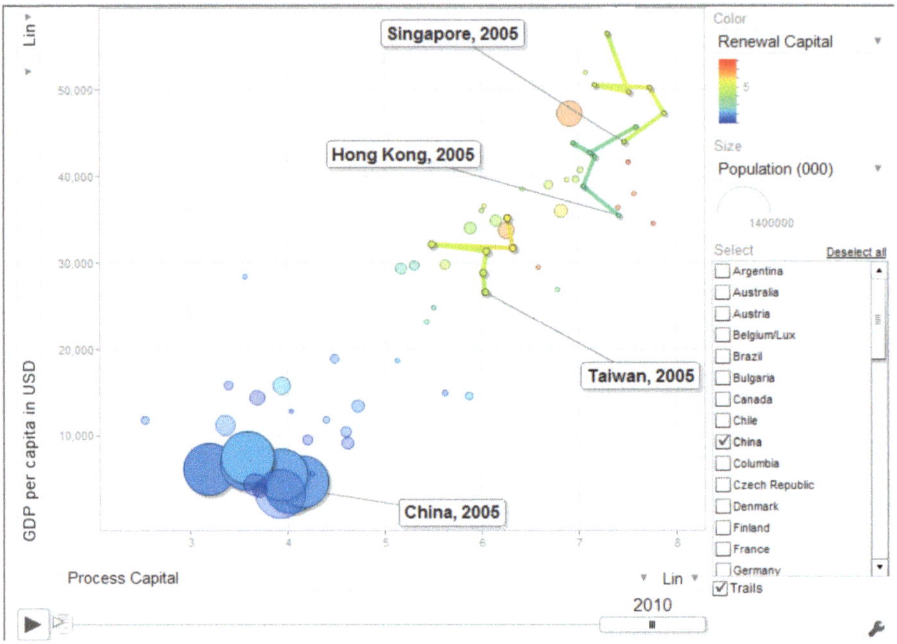

Fig. 3.11 The development of process capital and GDP per capita (ppp) for the Greater China economies, 2005–2010

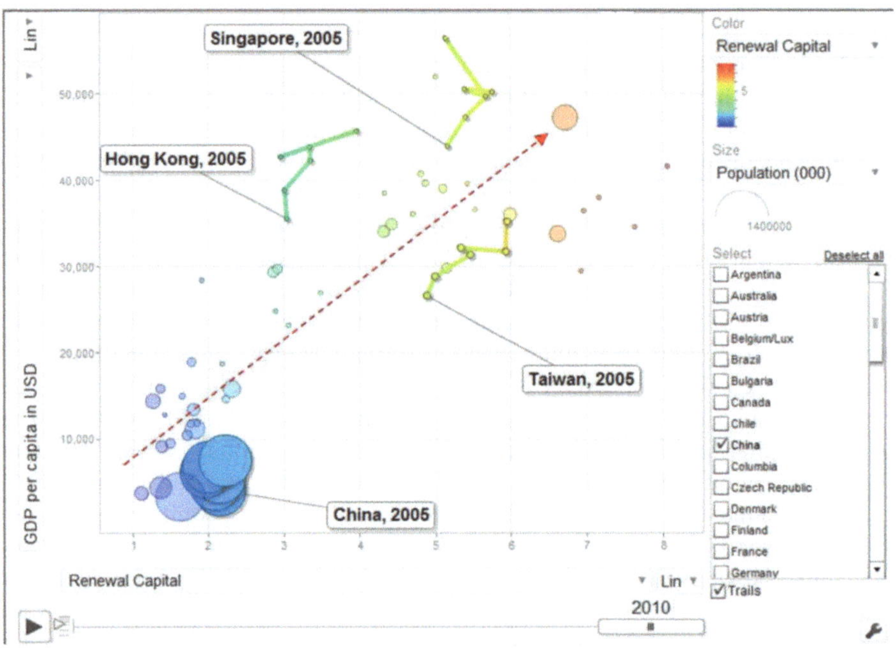

Fig. 3.12 The development of renewal capital and GDP per capita (ppp) for the Greater China economies, 2005–2010

Long-Term and Short-Term National Intellectual Capital

In addition to the six figures above that show the relationship of each individual capital with GDP, Figs. 3.13 to 3.16 further plot the relationships between human capital and renewal capital, as well as between market capital and process capital. In our first study (Lin and Edvinsson 2011: 324), we found that the combined increase in human capital and renewal capital would lead to a nation's long-term development, whereas the combined increase in market capital and process capital might result in more immediate economic development. To present their relationships, Figs. 3.13 to 3.16 map the four capitals on a coordinate with two types of presentation. Based on the same data, Figs. 3.13, 3.15 exhibit a scatterplot for the four economies only, whereas Figs. 3.14, 3.16 show all 48 countries in the background so that readers will know the relative positions of these four economies. Please note that the connation of color shade for Figs. 3.14, 3.16 has been changed from renewal capital to GDP per capital (ppp).

Both Figs. 3.13, 3.14 indicate that Singapore and Taiwan are at a similar high level and intertwined with each other in their long-term NIC development. Hong Kong is in between, and China still needs some time to catch up with the other three economies. Figure 3.14 further indicates that the Greater China economies possess long-term NIC in the lower cluster (China), the middle cluster (Hong Kong), and the middle upper cluster (Singapore and Taiwan) in the world map. If Singapore and Taiwan helped Hong Kong, and Hong Kong then pulled China along in developing these two long-term capitals, the potential value of the linked intangible assets of the Greater China economies would be powerful. These two capitals take time to develop and achieve, as human capital is based on unpredictable human performance and renewal capital is the end result of good human capital. Therefore, their development should be planned as early as possible.

Figures 3.15, 3.16 map market capital and process capital of the four economies on the same coordinate plane. Market capital refers to the general assets embodied in a nation's relationship with international markets while process capital refers to the national infrastructure, such as information systems and the number of days required to start a new business. In the short term, these are the two types of capital that can be expected to produce faster results if a nation is willing to engage in more extensive investments.

Figures 3.15 and 3.16 indicate that Hong Kong and Singapore are at a similar, relatively higher, level in terms of their short-term NIC, Taiwan is in the middle, and China is at the bottom corner of the graph. Again, Fig. 3.16 indicates that the Greater China economies possess short-term NIC in the lower cluster (China), the middle cluster (Taiwan), and the upper cluster (Hong Kong and Singapore) in this world map.

Using a supply chain analogy, the collaboration among these four economies can form a chain with a complete spectrum of services. In actuality, such collaboration has already been realized to a certain degree. For instance, manufacturing in China, logistic planning in Taiwan, and marketing in Hong Kong are becoming an effective

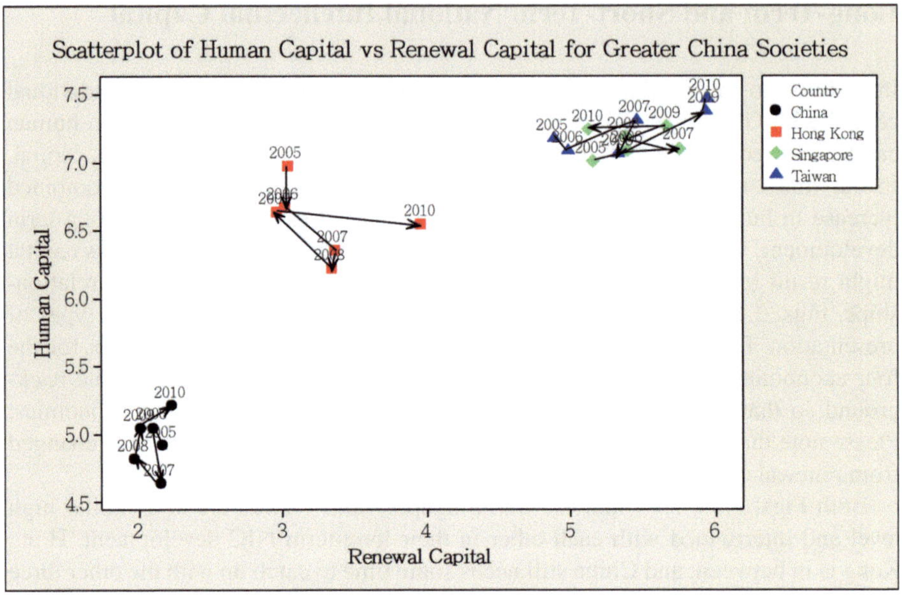

Fig. 3.13 Scatterplot of human capital versus renewal capital for China, Hong Kong, Singapore, and Taiwan

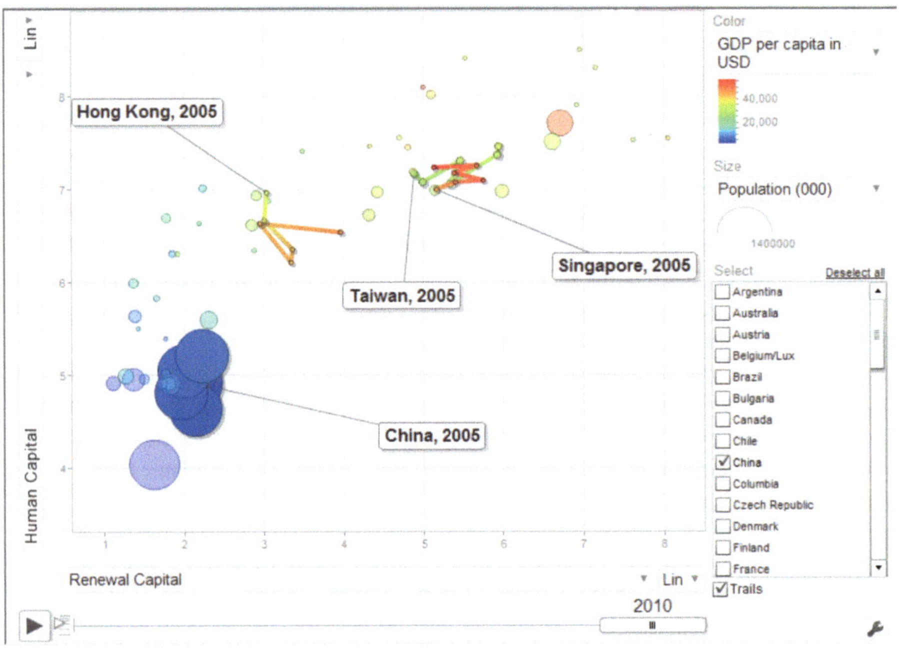

Fig. 3.14 Human capital versus renewal capital for China, Hong Kong, Singapore, and Taiwan

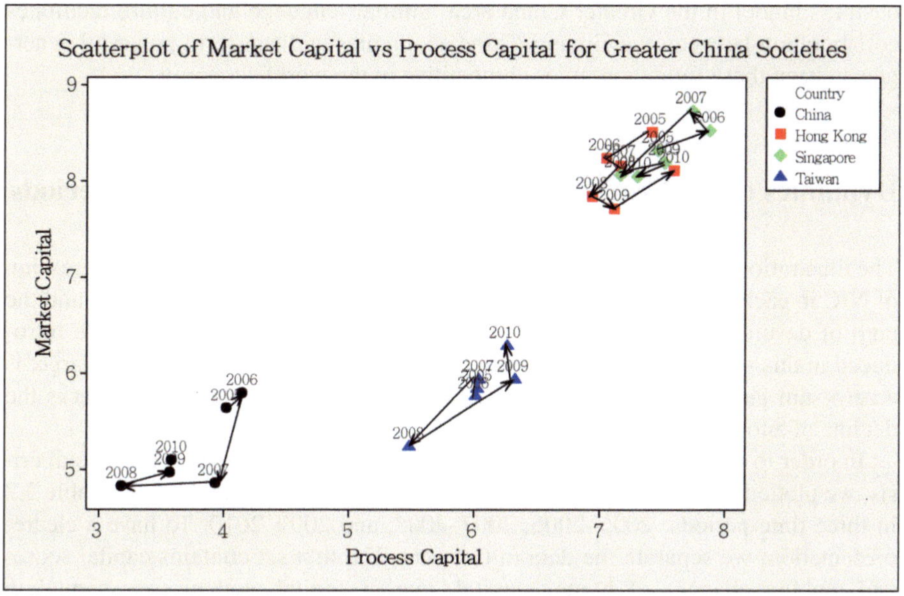

Fig. 3.15 Scatterplot of market capital versus process capital for China, Hong Kong, Singapore, and Taiwan

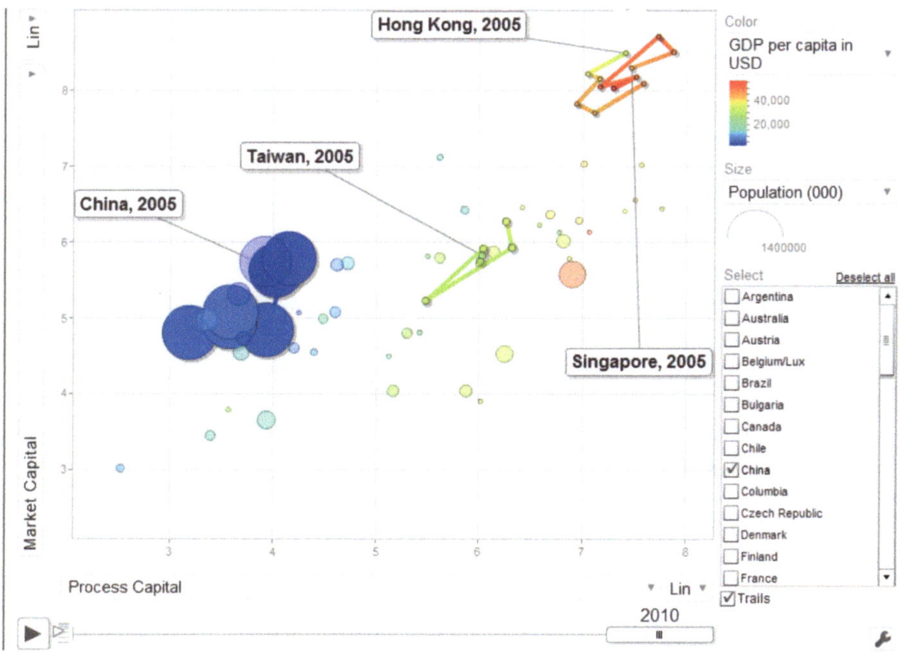

Fig. 3.16 Market capital versus process capital for China, Hong Kong, Singapore, and Taiwan

business model in the Greater China area. Similar language and culture facilitates collaboration between the Greater China economies and can form powerful synergies to keep these four economies competitive in the world economy.

Dynamics of National Intellectual Capital in Three Time Periods

The illustrations and graphs in the previous section describe the internal progression of NIC in each country as individual capital scores change over the 6 years and the path of development of various capitals. The ranking changes that will be introduced in this section represent external competitiveness. Rankings, by and large, is a zero-sum game in which the advance of one country simultaneously marks the decline of another.

In order to probe the dynamics of NIC change before and after the financial crisis, we plotted Figs. 3.17, 3.18, 3.19, 3.20, 3.21, 3.22, 3.23, and 3.24 and Table 3.2 in three time periods: 2005–2006, 2007–2008, and 2009–2010. To have a clearer presentation, we separate the data in two sets. The first set contains capital scores and ranking changes of human capital, market capital, and process capital in Figs. 3.17, 3.19, 3.21, and 3.23. The second set exhibits capital scores and ranking changes of renewal capital, financial capital, and average NIC in Figs. 3.18, 3.20, 3.22, and 3.24.

The figures are mapped based on the mean capital scores and ranking changes of each period. Ranking changes are next to each capital score and copied from the bold-faced ranking difference of columns 2, 5, 7, and 9 of Table 3.2. Taking human capital ranking difference of China (1, −4, 5, 0), for example the ranking change of “1,” representing the difference between the 3rd (2009–2010) and 1st (2005–2006) period, is plotted at 2005; the ranking change of “−4,” representing the difference between the 2nd period (2007–2008) and 1st (2005–2006), is plotted at 2007; the ranking change of “5,” representing the difference between the 3rd period (2009–2010) and 2nd (2007–2008), is plotted at 2009; and the ranking change of “0,” representing the difference between the 3rd period (2009–2010) and 6-year average ranking (exhibited in Table 3.1), is plotted at 2010. Special attention can be paid to the negative numbers, which indicate that the level of a particular capital was higher before than after the financial crisis. Since it is easier to read figures than tables, we will describe the findings of each figure separately.

As shown in Fig. 3.17 and Table 3.2, China’s market capital (−10, −9, −1, −1) and process capital (−9, −4, −5, −4) showed relatively large ranking declines, comparing 2009–2010 with 2005–2006 (plotted at 2005) and 2007–2008 with 2005–2006 (plotted at 2007). Negative ten (−10) and negative nine (−9) ranks in market capital means China was losing its market capital international competitiveness after and during the financial crisis. However, when comparing its ranking of 2009–2010 with that of the 6-year average (plotted at 2010), it dropped by only one rank.

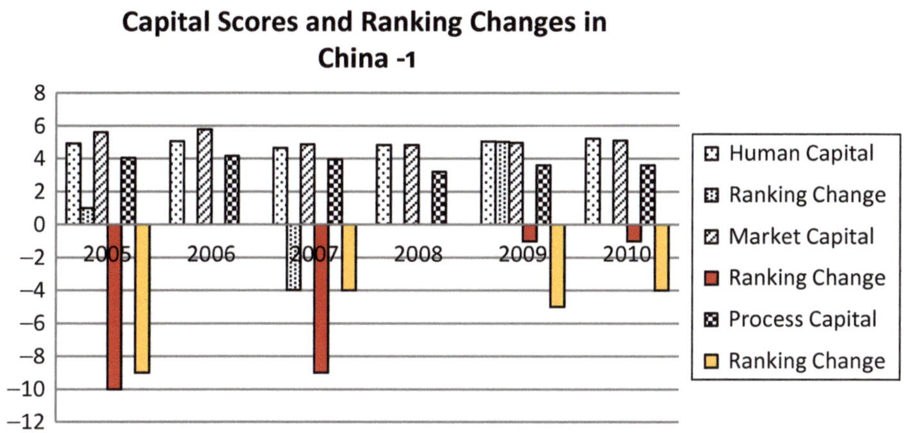

Fig. 3.17 Human capital, market capital, process capital, and ranking changes in China

China's process capital ranking declined in all four comparisons. Even though Figs. 3.3, 3.15 show that its score did not have much difference between the years 2007–2008 and 2009–2010, the ranking still dropped (plotted at 2009). This means, internally, there is not much difference, yet China is losing its international competitiveness in process capital. The last ranking drop of process capital (plotted at 2010) shows that when comparing its ranking of the most recent 2009–2010 period with that of the 6-year average, it declines by 4 ranks.

Figure 3.18 shows that China's renewal capital and financial capital do not experience large changes. Its NIC (−5, −7, 2, 2) shows a relatively large decline during and after the financial crisis; however, its most recent NIC (2009–2010) has improved two (2) ranks, comparing to that during the financial crisis (2007–2008) and comparing to that of its 6-year average.

Figure 3.19 shows that all four types of human capital ranking comparisons in Hong Kong declined, especially comparing 2009–2010 with 2005–2006 and 2007–2008 with 2005–2006. Market capital did not experience any ranking changes, and process capital had ranking advances. Figure 3.20 shows that Hong Kong stayed relatively steady in terms of renewal capital, financial capital, and average NIC international competitiveness.

Figure 3.21 shows that Singapore exhibited a slight decline in human capital ranking, no ranking changes in market capital, and a small decline in process capital. Figure 3.22 indicates that Singapore declined in renewal capital with little changes in financial capital and average NIC.

Figure 3.23 shows that Taiwan exhibited gains in market capital ranking (4, −2, 6, 3) with two rank (−2) declines in comparing 2007–2008 and 2005–2006 (plotted at 2007). This reflects the negative impact of the financial crisis. However, its six-rank (6) gain, in comparing 2009–2010 and 2007–2008 (plotted at 2009), showed

Table 3.2 Ranking changes in three time periods for China, Hong Kong, Singapore, and Taiwan

	(1st period–3rd period)	1st period 2005–2006	2nd period 2007–2008		3rd period 2009–2010		2005–2010 Average NIC48 ranking	
Human capital	Difference#1	Ranking	Ranking	Difference	Ranking	Difference	NIC 48 Ranking	Difference
China	**1**	40	44	**−4**	39	**5**	39	**0**
Hong Kong	**−7**	22	28	**−6**	29	**−1**	25	**−4**
Singapore	**−1**	16	15	**1**	17	**−2**	17	**0**
Taiwan	**3**	15	14	**1**	12	**2**	13	**1**
Market capital	Difference	Ranking	Ranking	Difference	Ranking	Difference	NIC 48 Ranking	Difference
China	**−10**	19	28	**−9**	29	**−1**	28	**−1**
Hong Kong	**0**	2	2	**0**	2	**0**	2	**0**
Singapore	**0**	1	1	**0**	1	**0**	1	**0**
Taiwan	**4**	18	20	**−2**	14	**6**	17	**3**
Process capital	Difference	Ranking	Ranking	Difference	Ranking	Difference	NIC 48 Ranking	Difference
China	**−9**	35	39	**−4**	44	**−5**	40	**−4**
Hong Kong	**4**	8	6	**2**	4	**2**	5	**1**
Singapore	**−1**	2	2	**0**	3	**−1**	2	**−1**
Taiwan	**4**	21	20	**1**	17	**3**	20	**3**
Renewal capital	Difference	Ranking	Ranking	Difference	Ranking	Difference	NIC 48 Ranking	Difference
China	**1**	32	33	**−1**	31	**2**	33	**2**
Hong Kong	**0**	23	22	**1**	23	**−1**	23	**0**
Singapore	**−2**	10	9	**1**	12	**−3**	11	**−1**
Taiwan	**6**	14	10	**4**	8	**2**	10	**2**
Financial capital	Difference	Ranking	Ranking	Difference	Ranking	Difference	NIC 48 Ranking	Difference
China	**1**	45	43	**2**	44	**−1**	44	**0**
Hong Kong	**1**	5	4	**1**	4	**0**	4	**0**
Singapore	**1**	2	2	**0**	1	**1**	2	**1**
Taiwan	**3**	22	19	**3**	19	**0**	20	**1**

NIC	Difference	Ranking	Ranking	Difference	Ranking	Difference	NIC 48 Ranking	Difference
China	**–5**	33	40	**–7**	38	**2**	40	**2**
Hong Kong	**0**	13	14	**–1**	13	**1**	14	**1**
Singapore	**1**	6	4	**2**	5	**–1**	5	**0**
Taiwan	**8**	19	17	**2**	11	**6**	18	**7**

#Remarks:

1. The second column compares the 3rd (2009–2010) period ranking with the 1st (2005–2006)
2. The fifth column compares the 2nd (2007–2008) period ranking with the 1st (2005–2006)
3. The seventh column compares the 3rd (2009–2010) period ranking with the 2nd (2007–2008)
4. The last column compares the most current 3rd period ranking (2009–2010) with the original 6-year average NIC48 ranking summarized in Table 3.1

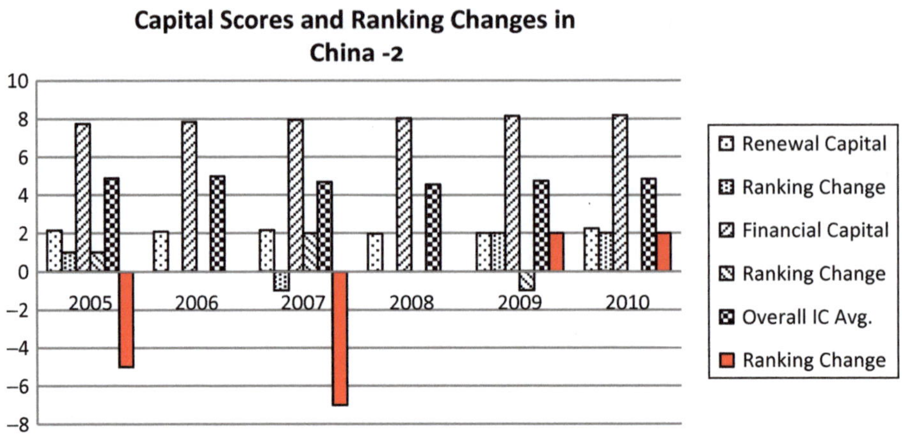

Fig. 3.18 Renewal capital, financial capital, average NIC, and ranking changes in China

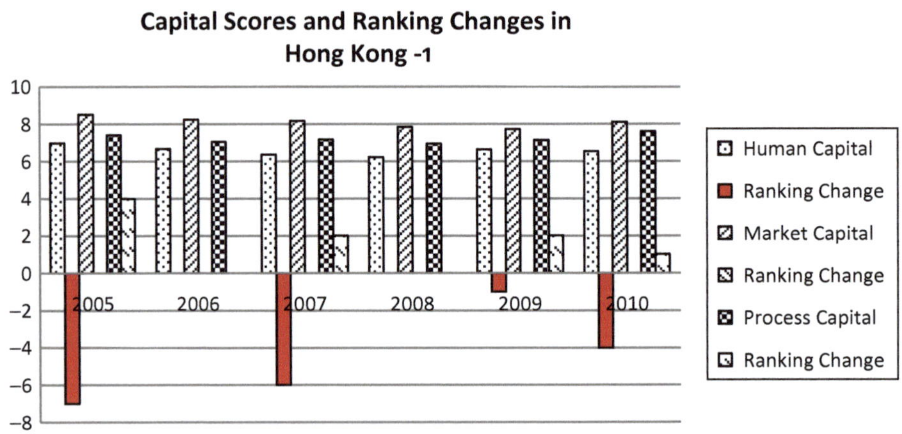

Fig. 3.19 Human capital, market capital, process capital, and ranking changes in Hong Kong

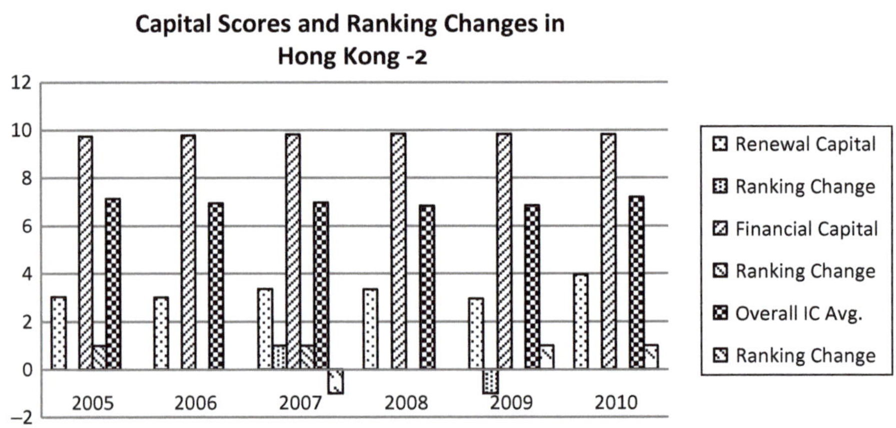

Fig. 3.20 Renewal capital, financial capital, average NIC, and ranking changes in Hong Kong

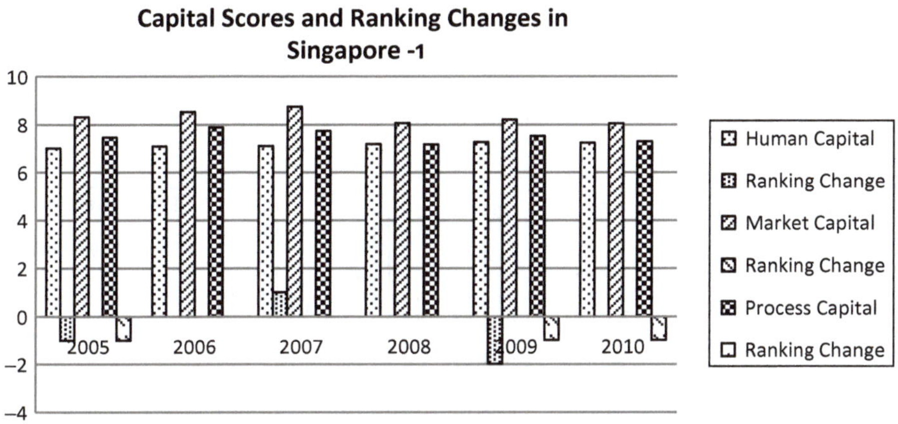

Fig. 3.21 Human capital, market capital, process capital, and ranking changes in Singapore

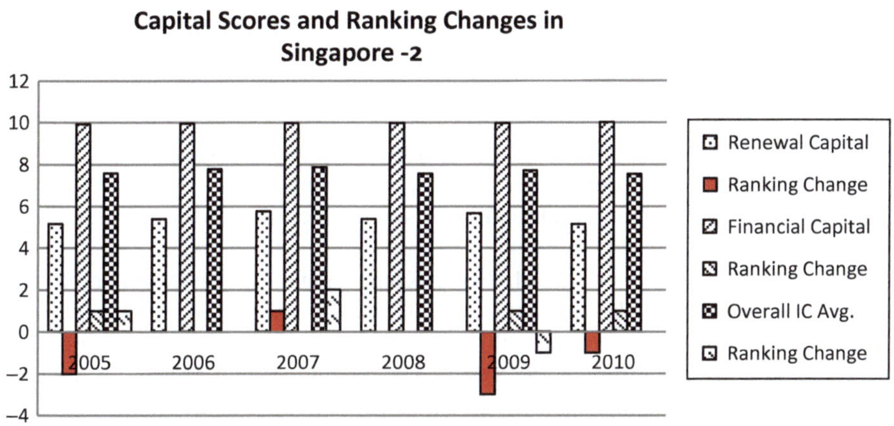

Fig. 3.22 Renewal capital, financial capital, average NIC, and ranking changes in Singapore

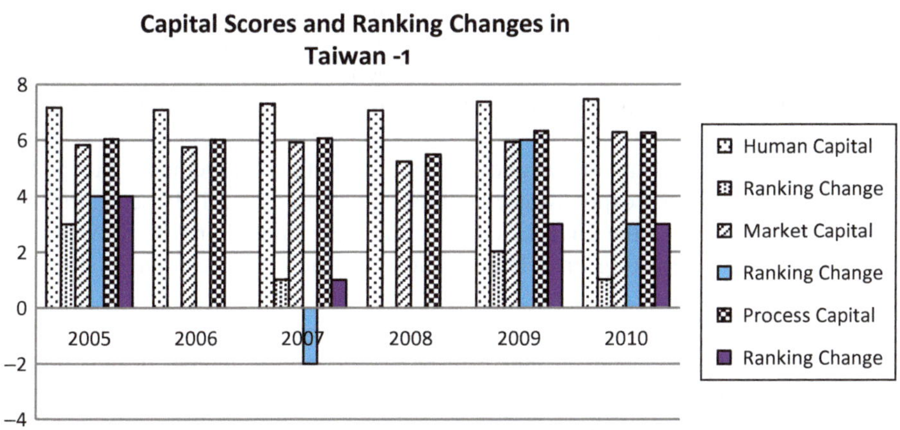

Fig. 3.23 Human capital, market capital, process capital, and ranking changes in Taiwan

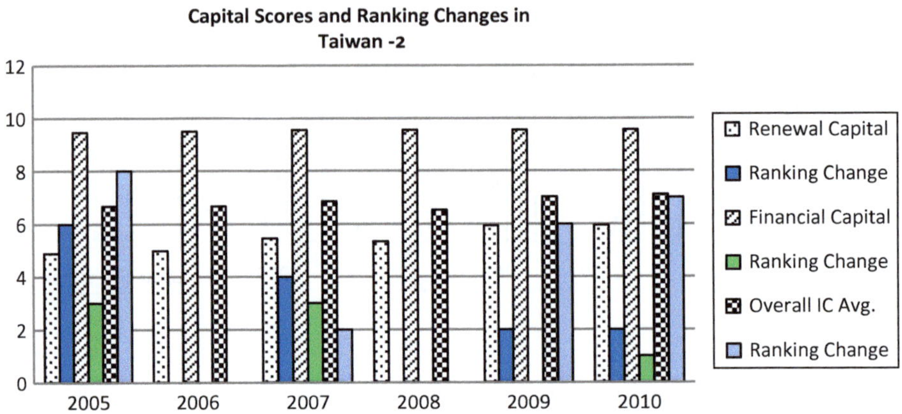

Fig. 3.24 Renewal capital, financial capital, average NIC, and ranking changes in Taiwan

a strong recovery after the financial crisis. Figure 3.24 also indicates that Taiwan has good ranking advancement in renewal capital, financial capital, and average NIC. Taiwan is the economy that had large scale rank gains in the Greater China economies.

In summary, for human capital, Hong Kong showed a downward trend and Taiwan an upward trend. For market capital, China's relatively high level of capital in the first period did not continue throughout the financial crisis. In addition, its market capital did not return to its 2005–2006 (#19 in Table 3.2) level of international competitiveness after the financial crisis (#29: 2009–2010). On the contrary, although Taiwan's market capital declined during the financial crisis (2007–2008) (#20), it picked up afterward (#14: 2009–2010) and reached higher levels than before the crisis (#18: 2005–2006).

For process capital, China declined in its international competitiveness after the financial crisis (#44: 2009–2010) and was not able to return to its 2005–2006 position (#35). Hong Kong and Taiwan have better process capital international competitiveness after the financial crisis with all positive ranking changes (Table 3.2).

For renewal capital, all the economies had ranking improvements except for Singapore. For financial capital, since the ranking is based on a 1–10 scale, the difference before and after the financial crisis is relatively small. For NIC, Taiwan was able to improve its ranking on a larger scale than the other economies after the financial crisis.

Taking a cursory look at Figs. 3.17, 3.18 through Figs. 3.23, 3.24, China has the largest number of rank declines, and Taiwan has the largest number of rank gains. That is, China is losing its NIC international competitiveness, whereas Taiwan is gaining in NIC international competitiveness post-financial crisis.

3-Dimensional National Intellectual Capital Trajectory

Up until this section, all the figures plotted different combinations of various types of capital, capital-GDP codevelopment, and ranking changes on a two-dimensional plane. Even though Figs. 3.7, 3.8, 3.9, 3.10, 3.11, and 3.12 displayed four dimensions with color and bubble size as the 3rd and 4th dimension in addition to x-axis and y-axis, graphically they are still on one plane. This section presents three-dimensional (3D) graphs showing each country's NIC development trajectory on a 48-country 3D landscape of various types. Since NIC evolution over time is always complex, combining different views from different angles may provide more comprehensive perspectives for analysis.

In what follows, we first explore the value-driving context by uncovering the path of GDP growth for China, Hong Kong, Singapore, and Taiwan from 2005 to 2010. Second, in explaining the GDP growth, we present the high, middle, and low capabilities which human capital, market capital, process capital, and renewal capital respectively have. Third, we examine the efficiency (GDP growth) enhancing factors and impeding factors of each economy, taking each economy's turning point into consideration. Fourth, we identify the efficiency drivers of each economy in reaching a targeted GDP.

Figure 3.25 shows a rectangular 3D landscape, with a longer north–south side. The formation of the landscape is based on the data of 29 national intellectual capital indicators (presented in Appendix 3) for 48 countries covering 6 years (2005–2010), with a total of 8,352 (29×48×6) data points. In other words, Fig. 3.25 is the aggregated NIC landscape of the 48 countries. The enlarged trails in Fig. 3.25 indicate NIC progression (upward), regression (downward), or configuration change (sideward) of each economy over the 6-year period. Each white trail on the landscape represents a country. The altitude coloring from purple, dark blue, blue, green, and yellow to red is computer generated, according to GDP values, with purple (north) representing high GDP and red (south) low GDP. Figure 3.26 presents the potential rotation or partial presentation of the 3D formation in order to provide readers a better view for the focal area of discussion.

The relative NIC positions of China, Hong Kong, Singapore, and Taiwan (pointed out with arrows) are located at different parts of the overall landscape. The NIC trails of Hong Kong and Singapore are located at the ridge, Taiwan's trail is at the upper left-hand side, and China's trail is at the lower central part of the 3D rectangular formation.

Figures 3.27, 3.28, and 3.29 form a set together. The areas of focus are shaded in light blue. Automated by the computer program to have a better view, the three figures are taken from front view toward the left side view as Fig. 3.26 indicates. Figure 3.27 presents four graphs indicating the high-capability region of human capital, market capital, process capital, and renewal capital in explaining the GDP growth. High capability means the data points are 20% above the mean value of each capital. This figure shows that both Singapore and Taiwan have all four capitals in the high-capability region. Hong Kong has three capitals (human capital, market capital, and process capital) in the high-capability region.

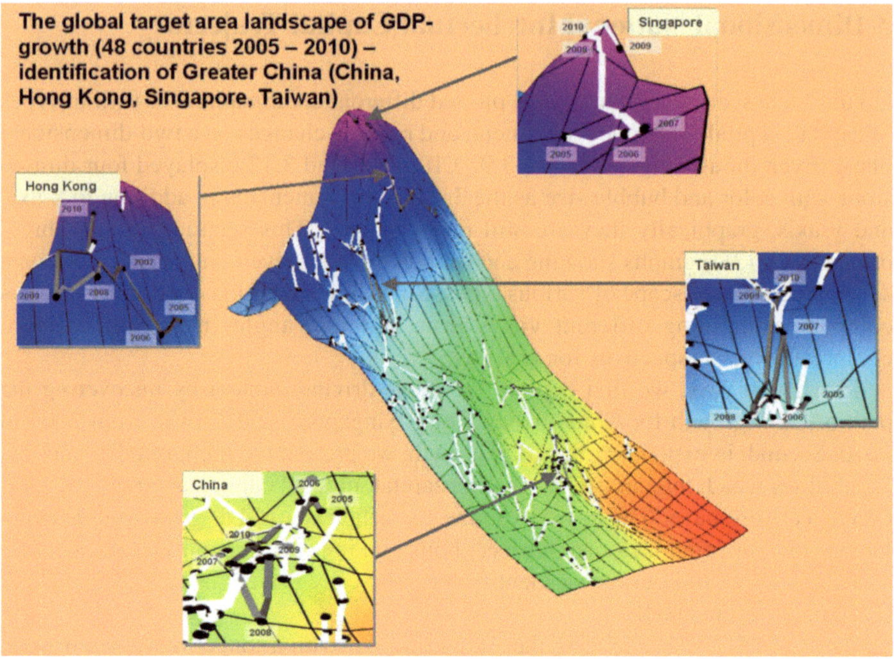

Fig. 3.25 The NIC trail of China, Hong Kong, Singapore, and Taiwan on a 3D 48-country landscape

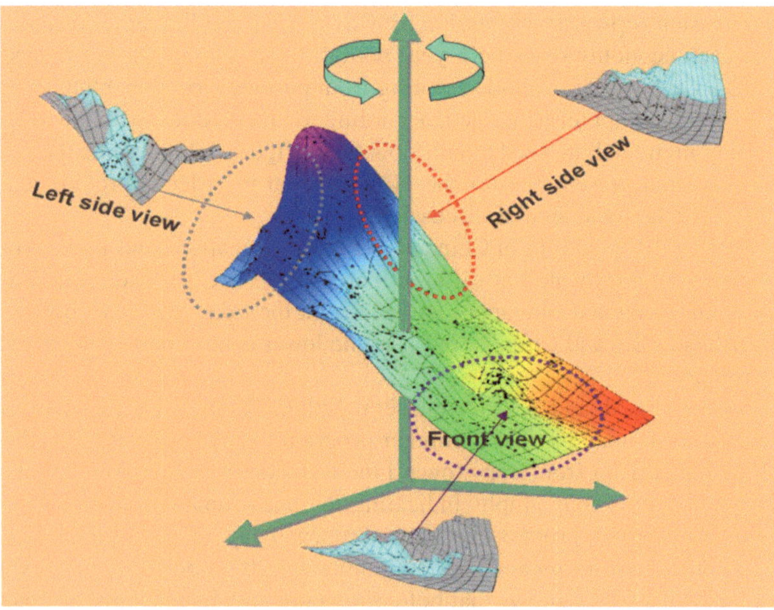

Fig. 3.26 The potential rotation and partial presentation of the 3D formation

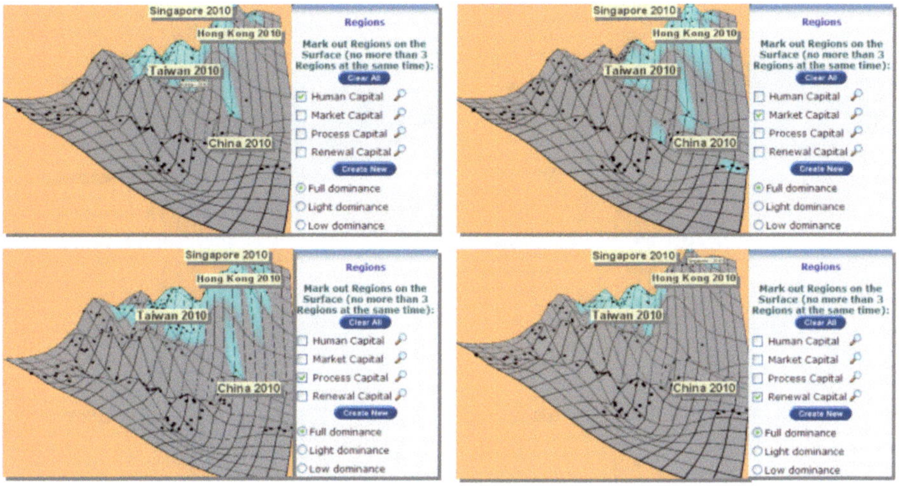

Fig. 3.27 The high-capability region of human capital, market capital, process capital, and renewal capital

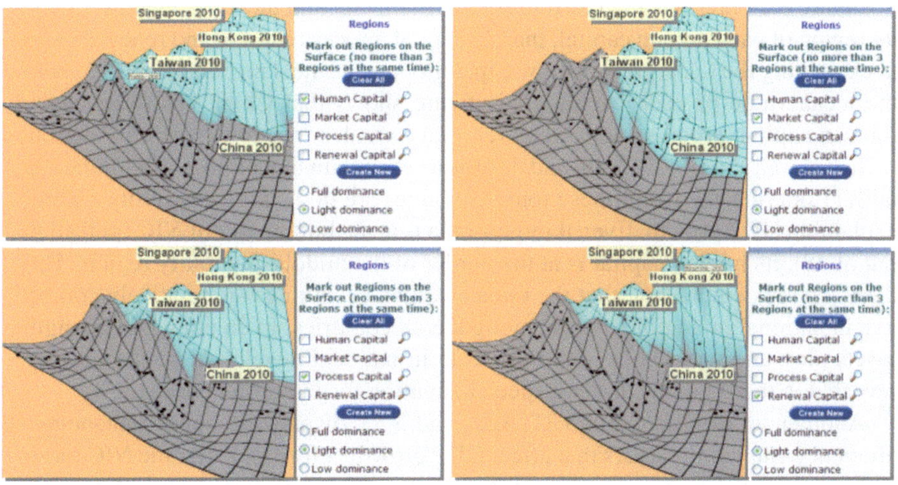

Fig. 3.28 The middle-capability region of human capital, market capital, process capital, and renewal capital

Figure 3.28 displays another four graphs indicating the middle-capability region of each human capital, market capital, process capital, and renewal capital in explaining the GDP growth. Middle capability means the data points are within the top 50% value of each capital. This figure shows that Hong Kong's renewal capital is in the middle-capability region, and China's market capital is at the borderline of middle-capability region.

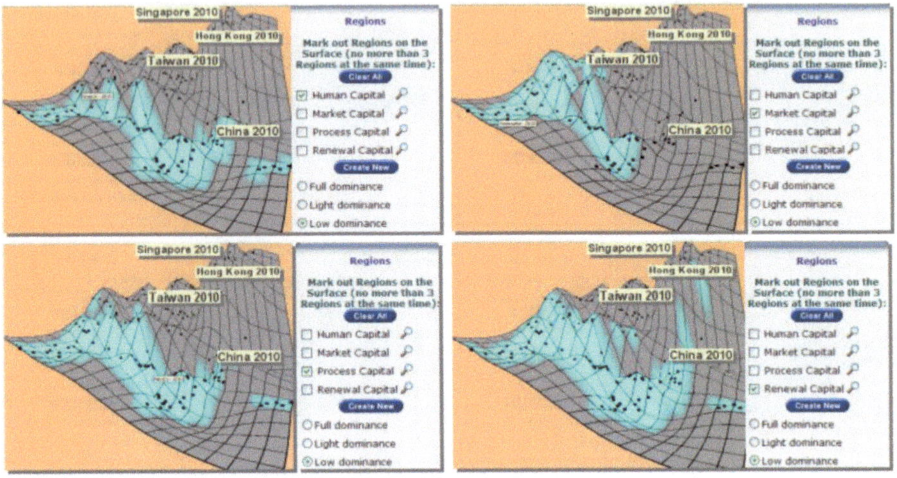

Fig. 3.29 The low-capability region of human capital, market capital, process capital, and renewal capital

Figure 3.29 presents the last four graphs in this set and indicates the low-capability region of each human capital, market capital, process capital, and renewal capital in explaining the GDP growth. Low capability means the data points are 20% below the mean value of each capital. This figure shows that China has three capitals (human capital, process capital, and renewal capital) in the low-capability region.

In general, Figs. 3.27, 3.28, and 3.29 as a set explain that almost 70% of the capitals (11 out of 4×4=16) of these four economies are in the high-capability region in explaining GDP growth. Even though China is at the bottom tier of NIC (ranked #40 out of 48), its market capital is at the border of the middle-capability region. These three figures demonstrate that the Greater China economies are relatively competitive in national intangible assets among the 48 countries. In an era when intangible assets play an increasingly important role in national development, the messages revealed in these graphs deserve special attention for interested parties.

Another set of Figs. 3.30, 3.31, 3.32, and 3.33, display efficiency-enhancing or efficiency-impeding factors that affect GDP growth before and after the NIC turning point for China, Hong Kong, Singapore, and Taiwan, respectively. The white indicator boxes at the upper right display the results of the enhancing or impeding factors. We employed three or four rounds of data analysis, depending on whether the country has one or two NIC turning points. For instance, China's turning points are 2007 and 2008; therefore, we traced the explaining factors of 2005 against 2007 (starting point → the first turning point), 2006 against 2007 (one year before the turning point → the first turning point), 2007 against 2008 (one year before the turning point → the second turning point), and 2008 against 2010 (the second turning point → the ending point).

As an easy reference, we summarize in Table 3.3 two sets of results, each with the five key factors. If there is a clear percentage difference cutting point (such as

Table 3.3 Enhancing factors and impeding factors of GDP growth for China, Hong Kong, Singapore, and Taiwan

	China	China	Hong Kong	Hong Kong	Singapore	Singapore	Taiwan	Taiwan
Influencing factor	Enhancing factors	Impeding factors	Enhancing factors	Impeding factors	Enhancing factors	Impeding factors	Enhancing factors	Impeding factors
Factor #1 before (1st) turning point	Capability availability (G)	Employee training (H)	Scientific articles (R&D)	Government efficiency (G)	R&D researchers (R&D)	Skilled labor (H)	Patents (R&D)	Mobile phone subscribers
Factor#1 after (2nd) turning point	*Employee training* (H)	*Higher education enrollment* (H)	*Scientific articles* (R&D)	*Patents* (R&D)	*R&D researchers* (R&D)	*Scientific articles* (R&D)	*Employee training* (H)	*Scientific articles* (R&D)
Factor #2 before (1st) turning point	Pupil-teacher ratio (H)	Openness of culture (M)	Computers in use per capita	Transparency of government policy (G)	Openness to culture (M)	Public expenditure on education (H)	R&D researchers (R&D)	Corporate tax encouragement (G)
Factor #2 after (2nd) turning point	*Enterprise—university cooperation* (R&D)	*Basic research* (R&D)	*Transparency of government policy* (G)	*Public expenditure on education* (H)	*Transparency of government policy* (G)	*Business R&D spending* (R&D)	*Corporate tax encouragement* (G)	*Higher education enrollment* (H)
Factor #3 before (1st) turning point	Enterprise—university cooperation (R&D)	Corporate tax encouragement (G)	Export of goods (M)	Globalization (M)	Employee training (H)	–	Higher education enrollment (H)	Convenience of establishing new firms (G)
Factor #3 after (2nd) turning point	*Internet subscribers*	*Pupil-teacher ratio* (H)	*Government efficiency* (G)	*Higher education enrollment* (H)	*Basic research* (R&D)	*Export of goods* (M)	*Capital availability* (G)	*Mobile phone subscribers*
Factor #4 before (1st) turning point	Internet subscribers	Basic research (R&D)	Literacy rate (H)	Basic research (R&D)	Basic research (R&D)	–	Scientific articles (R&D)	Openness to culture (M)

(continued)

Table 3.3 (continued)

	China	China	Hong Kong	Hong Kong	Singapore	Singapore	Taiwan	Taiwan
Influencing factor	Enhancing factors	Impeding factors	Enhancing factors	Impeding factors	Enhancing factors	Impeding factors	Enhancing factors	Impeding factors
Factor #4 after (2nd) turning point	*Capital availability* (G-C)	*Export of goods* (M)	*Skilled labor* (*H*)	*Image of country* (M)	*Patents* (R&D)	*Capital availability* (G-C)	*Openness to culture* (M)	-
Factor #5 before (1st) turning point	R&D % of GDP (R&D)	Image of country (M)	R&D researchers (R&D)	Openness of culture (M)	Scientific articles (R&D)	-	Export of goods (M)	Public expenditure on education (H)
	(Average weight—6.33 %) #	**(Average weight—22.83 %)** #	**(Average weight—17.21 %)**	**(Average weight—23.76 %)**	**(Average weight—13.12 %)**	**(Average weight—5.51 %)**	**(Average weight—9.11 %)**	**(Average weight—5.81 %)**
Factor #5 after (2nd) turning point	*Globalization* (M)	–	*Corporate tax encourage-ment* (G)	–	*Globalization* (M)	*Fair business competition environment* (G)	*Fair business competition environment* (G)	–
	(Average weight—18.43 %)	**(Average weight—3.99 %)**	**(Average weight—22.29 %)**	**(Average weight—10.38 %)**	**(Average weight—2.87 %)**	**(Average weight—13.25 %)**	**(Average weight—28.18 %)**	**(Average weight—9.44 %)**
Category frequency count	G—2 H—2 M—1 R&D—3	G—1 H—3 M—3 R&D—2	G—3 H—2 M—1 R&D—3	G—2 H—2 M—3 R&D—2	G—1 H—1 M—2 R&D—6	G—2 H—2 M—1 R&D—2	G—3 H—2 M—2 R&D—3	G—2 H—2 M—1 R&D—1

(G & G-C) government-related issues (total 16 counts including capital availability G-C 2 counts), *(H)* human capital issues (16 counts), *(M)* market capital issues (14 counts), *(R&D)* research and development (22 counts), # average weight of each enhancing/impeding factor before and after the turning point

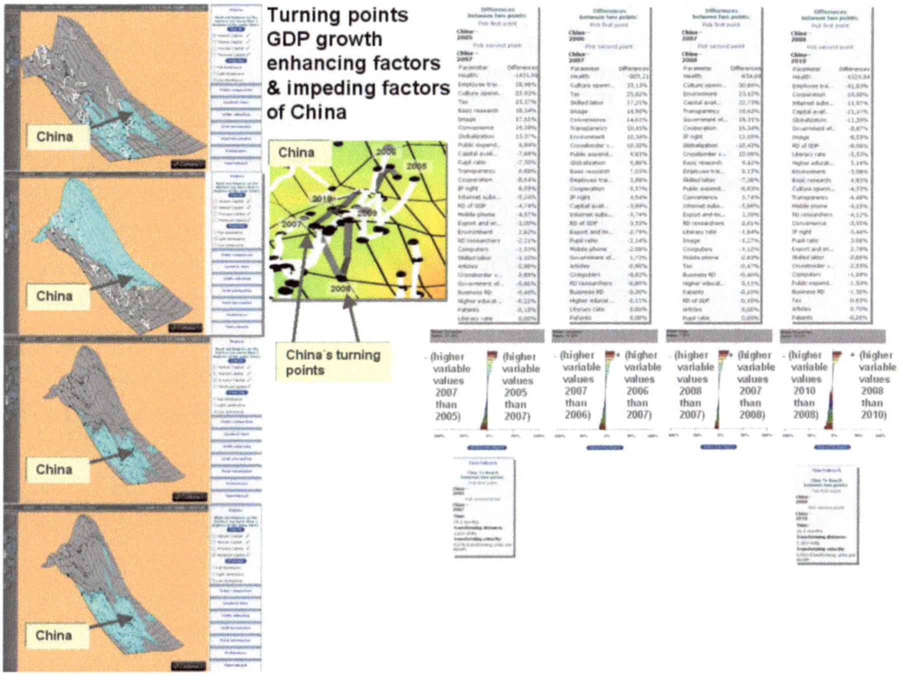

Fig. 3.30 Turning point and GDP growth enhancing and growth impeding factors of China

less than 1% in the white indicator boxes), the reported factors may be less than five. Results for the first set show the efficiency-enhancing and efficiency-impeding factors before the turning point. Results for the second set show the efficiency-enhancing and efficiency-impeding indicators after the (2nd) turning point in italic letters. Since the factor strength (percentage) varies with different explaining power, we have calculated the average weight of the five enhancing and impeding factors and display each respective percentage in the rows for "Factor #5." We also count the frequency of four more frequently displayed categories, namely, government-related issues (G), human capital-related issues (H), market capital-related issues (M), and research and development issues (R&D). Because capital availability was a concern during the financial crisis, we particularly highlighted this factor as "G-C."

Figure 3.30 shows the 2007 and 2008 turning points and the enhancing and impeding factors of China. The four rectangular 3D graphs on the left-hand side of the figure indicate that China's human capital, process capital, and renewal capital fall into the low-capability region (20% below average), yet its market capital is on the border of the middle-capability region.

Table 3.3 shows that in China, "capital availability" is the first enhancing factor before its first turning point and the fourth enhancing factor after its second turning point. That is, capital availability is not a problem for China before and after the

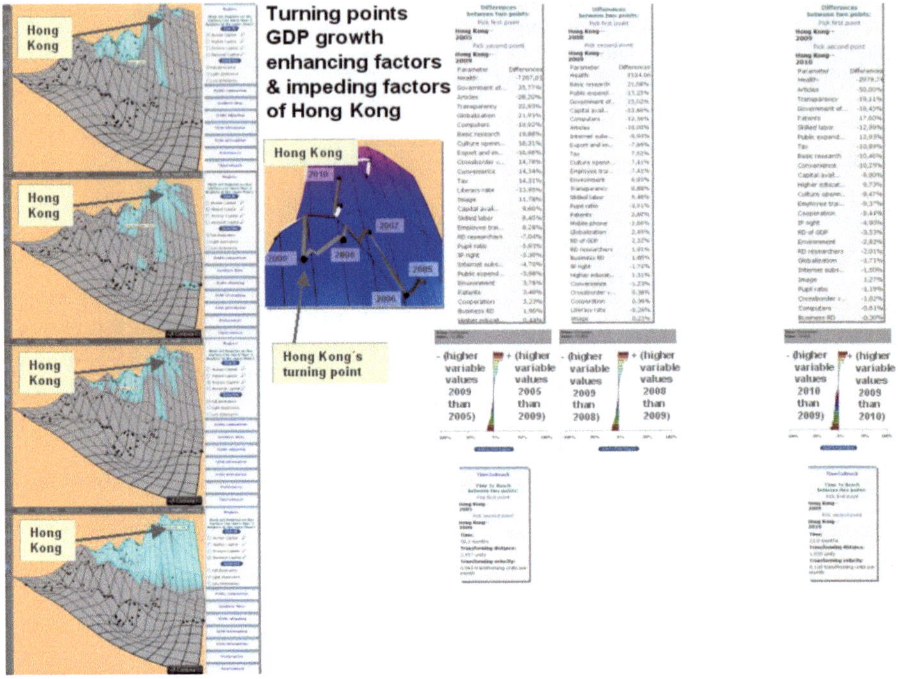

Fig. 3.31 Turning point and GDP growth enhancing and growth impeding factors of Hong Kong

financial crisis. A good sign in China is that employee training improved from the first impeding factor before the turning point to the first enhancing factor after the second turning point. That is, China's employee training greatly improved after the financial crisis and enhanced its GDP. Yet, its pupil-teacher ratio has been changed from the second enhancing factor before the turning point to the third impeding factor after the second turning point. Basic research continues to be an impeding factor before and after the turning point. Other impeding factors, such as higher education enrollment, openness of culture, corporate tax encouragement, export of goods, and image of country also deserve some attention.

In Fig. 3.31, Hong Kong's turning point is 2009. The four rectangular 3D graphs on the left-hand side indicate that its human capital, market capital, and process capital are in the high-capability region (20% above average), and renewal capital is in the middle-capability region (top 50%). Table 3.3 shows that scientific articles remains Hong Kong's first enhancing factor before and after the turning point. In addition, Hong Kong has two improvements, government efficiency changes from the first impeding factor before the turning point to the third enhancing factor after the turning point and transparency of government policy changes from the second impeding factor before the turning point to the second enhancing factor after the turning point. Other impeding factors, such as patents, public expenditure on

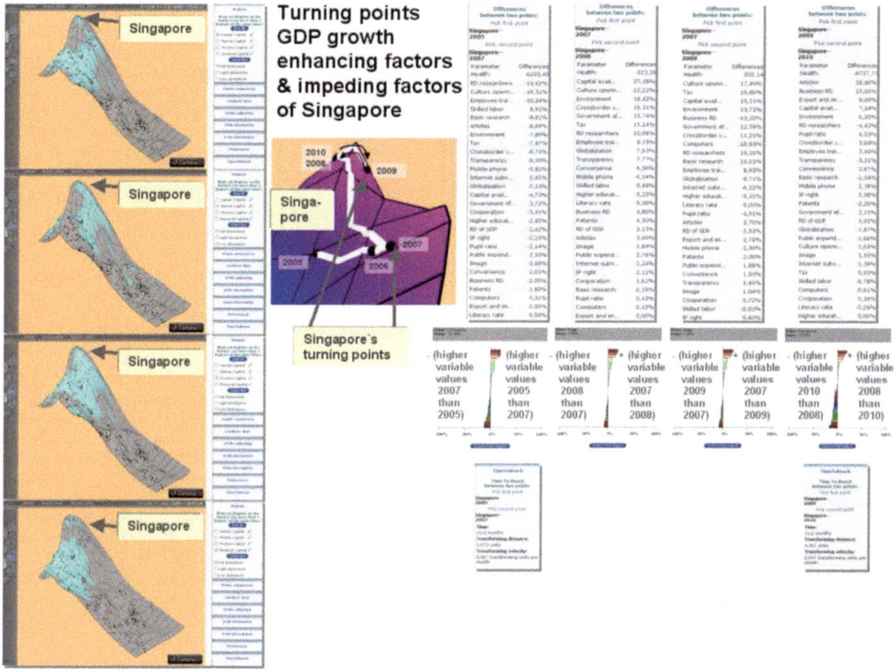

Fig. 3.32 Turning point and GDP growth enhancing and growth impeding factors of Singapore

education, globalization, higher education enrollment, basic research, image of country, and openness of culture also deserve some attention.

Figure 3.32 shows two turning points for Singapore in 2007 and 2009, respectively. The four rectangular 3D graphs on the left-hand side of Fig. 3.32 indicate that all the four capitals of Singapore are in the high-capability region (20% above average). Table 3.3 shows that R&D researchers remains the first enhancing factor before and after the turning point, and basic research is the fourth enhancing factor before the turning point and the third enhancing factor after the turning point. Yet, scientific articles changes from the fifth enhancing factor before the turning point to the first impeding factor after the turning point. Other impeding factors, such as skilled labor, public expenditure on education, business R&D spending, export of goods, capital availability, and fair business competition environment also deserve some attention. Interestingly, capital availability of Singapore became a concern after the financial crisis, which somewhat reflects its rising general government debt (Fig. 2.2).

Figure 3.33 shows that 2007 and 2008 are Taiwan's two turning points. The four rectangular 3D graphs on the left-hand side of Fig. 3.33 indicate that all four capitals of Taiwan are in the high-capability region (20% above average). Table 3.3 shows that R&D researchers are the first enhancing factor before the turning point, which partially reflects Taiwan's number one rank in the world for utility patents

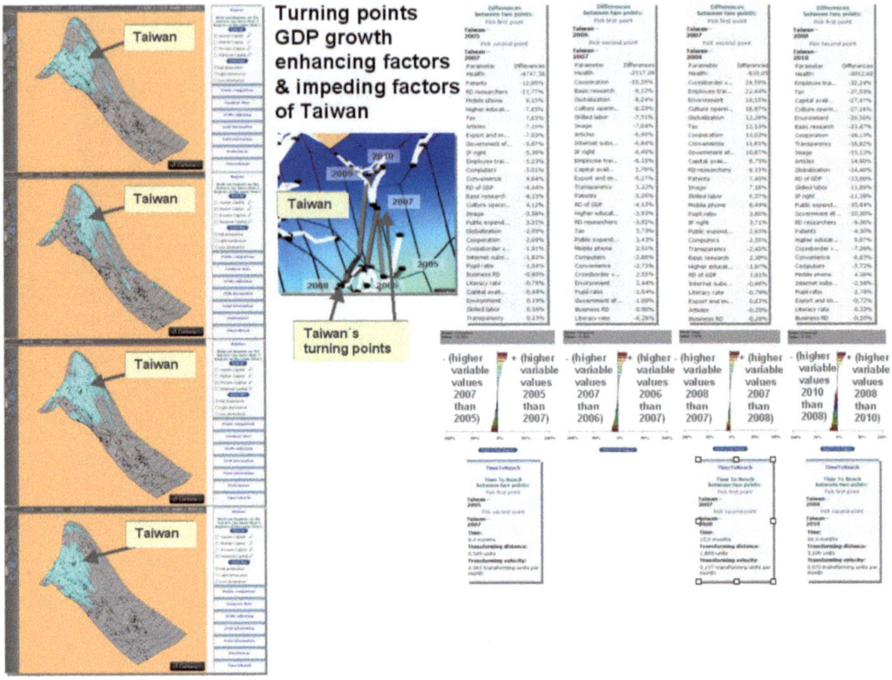

Fig. 3.33 Turning point and GDP growth enhancing and growth impeding factors of Taiwan

(Schwab 2011). Employee training as the first enhancing factor after the turning point can somewhat explains the effect of Taiwan's stimulus package by conducting government-sponsored employee training during the financial crisis. For Taiwan, corporate tax encouragement improves from the second impeding factor before the turning point to the second enhancing factor after the turning point, and openness of culture improves from the fourth impeding factor before the turning point to the fourth enhancing factor after the turning point. However, higher education enrollment changes from the second enhancing factor before the turning point to the second impeding factor after the turning point, and scientific articles changes from the third enhancing factor before the turning point to the first impeding factor after the turning point. Other impeding factors, such as mobile phone subscribers and convenience of establishing new firms, also deserve some attention.

To summarize Table 3.3 into an overall suggestion, we count the frequency of both enhancing and impeding factors by categories for all the four economies, shown in the last row of the table. There are a total of 16 counts of government-related issues (G), including 2 counts of capital availability. Among which, there are 7 counts of impeding factors, including capital availability, transparency of government policy, government efficiency, corporate tax encouragement, fair business competition environment, and convenience of establishing new firms. Hong Kong, Singapore, and Taiwan each has two counts of government-related issues as the impeding factors.

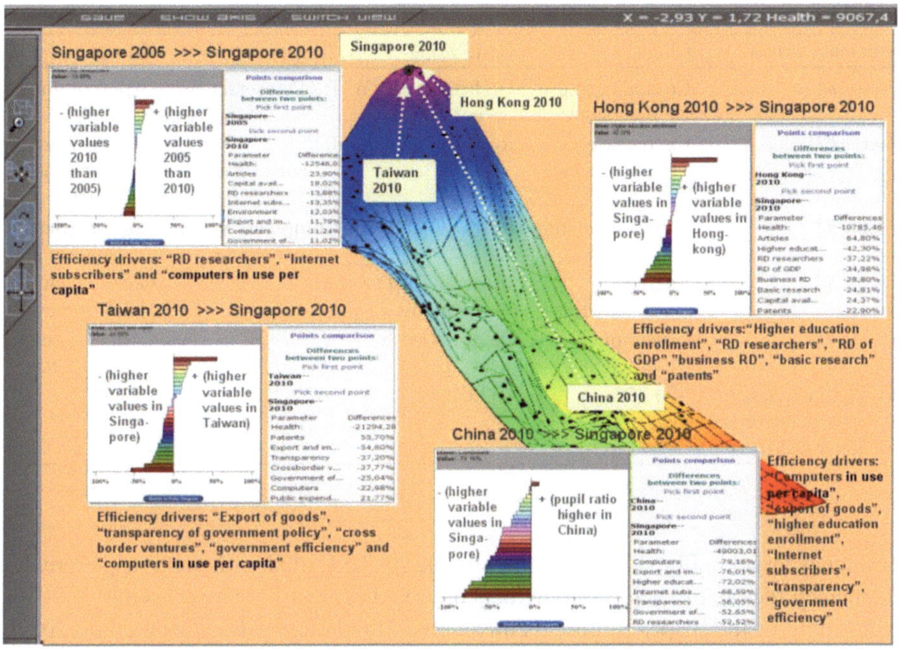

Fig. 3.34 Efficiency drivers and distance to targeted GDP of the Singapore

There are 16 counts of human capital issues (H). Among which, there are 9 counts of impeding factors, including employee training, higher education enrollment, pupil-teacher ratio, public expenditure on education, and skilled labor. There are 14 counts of market capital issues (M). Among these, there are 8 counts of impeding factors, including globalization, openness of culture, image of country, and export of goods. For research and development issues (R&D), there are 22 counts, among which impeding factors total 7 counts, including business R&D spending, basic research, patents, and scientific articles.

For impeding factors, the average weight before the turning point is higher than after the turning point in China (22.83% vs. 3.99%) and Hong Kong (23.76% vs. 10.38%). This result reflects the fact that the impeding factors for these two economies were less stressful after the financial crisis. On the contrary, the average weight before the turning point is lower than after the turning point in Singapore (5.51% vs. 13.25%) and Taiwan (5.81% vs. 9.44%), indicating that the impeding factors after the turning point deserve some attention.

Figure 3.34 identifies the efficiency drivers (positive term) for each individual economy. Efficiency targeting is like benchmarking. Efficiency represents the GDP in this 3D software program. Since Singapore is ranked number two in financial capital among the 48 countries (surpassed by Norway only), it is already the best benchmarking target for the other three Greater China economies. Therefore, Singapore's GDP is the

Table 3.4 The first five efficiency drivers targeting GDP of Singapore

Efficiency driver	China	Hong Kong	Singapore	Taiwan
Factor #1	Computer in use per capita	Higher education enrollment	R&D researchers	Export of goods
Factor #2	Export of goods	R&D researchers	Internet subscribers	Transparency of government policy
Factor #3	Higher education enrollment	R&D % GDP	Computers in use per capita	Cross-border ventures
Factor #4	Internet subscribers	Business R&D spending	-	Government efficiency
Factor #5	Government efficiency	Basic research	-	Computer in use per capita
Average deficiency percentage of the above 5 drivers	−70.37%	−33.62%	−12.82%	−33.52%

efficiency target in Fig. 3.34. In this figure, the stacked small horizontal stripes are the deficiency factors of each economy benchmarked against Singapore's GDP in 2010. The length of the stripes represents the degree of deficiency. For easier comparison, we list the first five deficiency factors of each economy in Table 3.4.

The results indicate that China needs to strengthen its "computer in use per capita," "export of goods," "higher education enrollment," "Internet subscribers," and "government efficiency." Hong Kong needs to strengthen its "higher education enrollment," "R&D researchers," "R&D % of GDP," "business R&D spending," and "basic research." Singapore still needs to strengthen its "R&D researchers," "Internet subscribers," and "computers in use per capita." Taiwan needs to strengthen its "export of goods," "transparency of government policy," "cross-border ventures," "government efficiency," and "computer in use per capita."

With Singapore being used as the benchmarking country, irrespective to the sequence of deficiency, the common areas (in frequency count) that these four economies need to improve include computers in use per capita (3 counts), higher education enrollment (2 counts), Internet subscribers (2 counts), export of goods (2 counts), government efficiency (2 counts), and R&D researchers (2 counts). From Table 3.4, the areas China needs to improve are basic issues, such as computers in use per capita, higher education enrollment, and Internet subscribers. Hong Kong needs to enhance its R&D-related capability, such as R&D researchers, R&D percentage of GDP, and business R&D spending. Although Singapore benchmarks itself, its weakest area is R&D researchers. Taiwan needs to enhance its government performance more, such as transparency of government policy and government efficiency. Based on the average deficiency (negative term) percentage reported in Table 3.4, China has the longest distance to achieve the target GDP (Singapore's), followed by Hong Kong and Taiwan. Hong Kong and Taiwan have very similar distances (−33.62% and −33.52%, respectively) to the target GDP, yet with different (R&D vs. government) concerns.

The added value of the above 3D trajectory analysis is to (1) display the NIC trails of the discussed countries on a 48-country landscape, (2) locate the capability region of the four individual capitals for each country, (3) highlight NIC turning points, details of the enhancing or impeding factors, and the dynamics of factor changes for each country, and (4) calculate the distance that each country has to progress in order to achieve the benchmark GDP and identify the first five key factors to reach this goal. This section supplements the previous sections by looking into the contribution of each individual indicator in each year, rather than from the perspective of average scores and average rankings.

In this chapter, we reported our research findings in layman terms with graphs and tables of various types for easier reading. This study on national intangible assets has uncovered some important findings that are not apparent in other GDP-related studies. For instance, the average distance of the first five efficiency drivers to the same targeted GDP for Hong Kong (–33.62%) is very similar to that of Taiwan (–33.52%), even though most of the figures indicate that Hong Kong has a much better performance than that of Taiwan.

Chapter 4
Beyond the 2008 Global Financial Crisis

This chapter reports activities after 2010, a timeline regarded as the aftermath of the 2008 global financial crisis which was formally announced to be over at the end of 2009 (Kehoe 2010; OECD 2010). Although some European countries were still in sovereign debt trouble after 2010 (Greece requested another bailout in early 2012 and Spain asked for external financial assistance in June 2012), we use 2010 as a cutoff point in order to be consistent with the reports of other country clusters in this booklet series. From the statistics introduced in Chapter 3, situations of these four economies in 2010 have clear rebounds under the support of each individual economy's stimulus efforts and the gradual revival of global demands.

As of mid-2012, the Asian economy as a whole has recovered from the financial crisis. The chief of the Asian Development Bank said, "The global economic slowdown will affect Asia this year (2012) but the region will remain the economic powerhouse of the world, led by China, India and Indonesia…regional and domestic demand in Asia are still fairly robust" (Kuroda 2012). At the same time, the IMF was gloomy about the world economy and cut its forecast for global economic growth for 2012 to 3.3% from 4%, because of the likelihood of a mild recession in Europe (Kuroda 2012). Affected by European market slowdown—China's major export destination, there was a prediction that China's growth would slip to 8.4% in 2012 (Debnath 2012). However, Asia's economic development relies on China more and more as it has become the top export destination for many of its neighbors, such as the other three economies reported in this volume.

In what follows, in the sequence of China, Hong Kong, Singapore, and Taiwan, we briefly describe what had happened in each individual economy after 2010 with some prior reasons.

C.Y.-Y. Lin et al., *National Intellectual Capital and the Financial Crisis in China, Hong Kong, Singapore, and Taiwan*, SpringerBriefs in Economics 8,
DOI 10.1007/978-1-4614-5984-2_4, © The Author(s) 2013

China

According to OECD, in 2010, the world bounced back from the worst recession since the Great Depression, and growth was once again close to 5%, of which China contributed almost one third of global growth (Herd et al. 2011). In 2010, China's GDP was valued at US$5.87 trillion, surpassing Japan's US$5.47 trillion to become the world's second largest economy after the US Standard Chartered predicts that China will become the world's largest economy (nominal GDP) by 2020 (Adam 2010). Similarly, US News (2008) reported that this will happen by 2035. Furthermore, China is now the largest exporter and second largest trader in the world while also being the largest destination for foreign direct investment and the fourth largest outward foreign direct investment among developing countries (OECD 2008). In addition to these promising prospects, China announced its 12th Five-Year Plan in March 2011, and the government vowed to continue reforming the economy with an emphasis on increasing domestic consumption in order to make its economy less dependent on exports for GDP growth in the future (Economy Watch 2010a).

Although impacted by this wave of financial crisis, China did not suffer a financial sector meltdown, credit crunch, or household debt crisis. China held the world's biggest foreign exchange reserves with US$1.53 trillion at the end of 2007 and had one of the highest gross national savings rates (50% of GDP in 2007) (Heilmann and Schmidt 2010). As a result, consumers were able to resume spending as soon as confidence returned. The September 2009 issue of the *China Retail Quarterly* reported not only strengthening consumption but also the fact that growth is coming from consumers outside of the marquee cities (Beijing, Shanghai, and Guangzhou) as the masses of rural China were the main source of domestic retail (Fleet 2010). In addition, urban retail sales for the first half of 2009 grew 13.7% year on year even in the midst of the global crisis, whereas rural sales grew by 37.6%, according to China's Directorate General of Budget, Accounting, and Statistics.

In 2011, China continues to run a very large trade surplus. New construction soared with a government campaign to provide more affordable housing (Bradsher 2011). As the crisis receded, IMF reported that China's ongoing withdrawal of monetary stimulus and the scaling back of public spending were leading to a healthy slowdown. Despite this slightly slower pace, the IMF also predicted that China's growth will continue to propel the world economy forward, contributing as much as one-quarter of the overall global growth of 4%. With the heightened uncertainty in the United States and Europe in 2011, China's exports still maintained strong momentum, growing 25% in August 2011 from a year earlier (IMF 2011). Import growth has also remained strong through commodities and capital goods. Foreign exchange reserves continue to climb, growing by a further US$153 billion in the second quarter of 2011 to US$3.2 trillion (IMF 2011). To reduce over reliance on the U.S. and EU markets, China was also increasingly diversifying export destinations toward other emerging markets.

Furthermore, in order to construct a balanced economy that relies more on domestic consumption and directs more resources to non-tradable sectors (such as retail, health, education, and other service industries), China continues to build out its pension system, pursue the comprehensive healthcare reform, and increase spending in areas that boost productivity in agriculture and non-tradable sectors.

However, China's continuous booms and extra cash have sent inflation at the consumer level surging to more than 6%. To fight inflation, in 2011 the Chinese financial authorities had set strict administrative quotas on new loans, limiting both the banks' ability to lend and businesses' ability to borrow (Bradsher 2011). Adding in China's planned slowdown, fighting for inflation, the pressure to appreciate its currency RMB from the U.S., and the decreasing demands from the advanced countries, in 2011 the country's huge manufacturing sector was starting to slow and orders were weakening, especially for exports. A question to be answered is whether China's slower economic growth and continued inflation could lead to more social tensions, which in turn could damage the economy through weaker investment and more cautious spending (Bradsher 2011).

After the financial crisis, there are two major economic challenges facing China—inflation and local government debt. In late 2010, even with deliberate suppression, China's inflation was still surpassed the government's target of 3%. In addition, local government debt swelled due to the off-the-books stimulus policies (Economy Watch 2010a).

Hong Kong

Since China has long been Hong Kong's largest trading partner, accounting for about half of Hong Kong's exports by value, China's fast recovery and strong economic performance also preserved Hong Kong from serious external shocks during the financial crisis. This then was a key factor in Hong Kong's speedy recovery. In 2010, Chinese companies constituted about 19% of the firms listed on the Hong Kong Stock Exchange and accounted for 62% of the Exchange's market capitalization (CIA 2012). Hong Kong has also established itself as the premier stock market for Chinese firms seeking to list abroad. In addition, Beijing's initial steps toward an internationalization of its RMB currency opened up new opportunities for Hong Kong's banks as well.

Hong Kong's economy has appeared stronger than expected after the financial crisis, growing 7.1% in the first three quarters of 2010 (Economic Forum 2010). Hong Kong business and other services had already shown a marked revival since the later part of 2009, benefiting from the increase of China-related IPO activities and the return of business in tandem with the regional recovery (HK government 2009). Although trades to the advanced economies slowed, exports to Asia and China, in particular, grew strongly in 2010 with the further recovery in Asia and vibrant growth in Mainland China. The continued relief measures announced during the financial crisis and Hong Kong's 2010/11 budget will boost its consumption spending (HK government 2009). In addition, after China's easing of travel restrictions to Hong Kong (CIA 2012), visitors from China accounted for 60% of the total visitors arriving in Hong Kong. This figure reached 23 million in 2010, which is a sharp increase of 26% from the previous year (Economic Forum 2011). The continuous spending of visitors from China has brought a continuous macroeconomy improvement in Hong Kong.

In spite of a promising future, the impacts of the 2008 financial crisis upon Hong Kong mainly resulted in two potential constraints (Tuan and Ng 2011). First, with a major share of foreign-owned financial sector, the global trend of tightening regulations on institutional governance, operations, and innovations in products may lead to the contractions in business value of the Hong Kong financial market. Second, increasing protectionism in commodity and service trade is expected for China's major trading partners, which may have spillover effects on Hong Kong. Furthermore, as an export dependent economy, Hong Kong consumers' confidence dipped a little in 2011 due to the fact that U.S. consumers are still struggling with high unemployment. In addition, the sovereign debt issue in Europe also remains a point of concern as well as raw material prices increase (Economic Forum 2010; 2011).

Yet, opportunities also surface with Hong Kong's unique position as China's gateway and the most internationally competitive city in the world. With the advantage of being permitted to take "different (political/economic) system" from China, Hong Kong is becoming the top international financial center of China. Its role is to direct China's international investment with the global world and help China integrate with the international financial market. In addition, the rapid growth in China's domestic market, expansion in private consumption, and the rise of private smaller enterprises have fueled the need for Hong Kong's services. Main areas of required service include infrastructural supports in transportation, environment protection, health care, social welfare, and education. Furthermore, the demand for improving business operations in China, such as market competition policies and SMEs or industrial policies, is also on the rise. These are areas which can further spur Hong Kong's growth in the future.

Throughout its history, Hong Kong has proven to provide a business-friendly investment environment and an industrious, quality labor force and human capital with pragmatic mentality and vital entrepreneurship (Tuan and Ng 2011). After the financial crisis, Hong Kong's human capital, market capital, process capital, and renewal capital can be put to use to help China prosper and integrate with the world economy through its participation and utilization of China's 12th "Five-Year Planning." Particularly, its reduced transaction cost due to political and economic autonomy from China provides ample opportunities for Hong Kong to become a key player in the Greater China economies in the years to come.

Singapore

Heavy reliance on exports hurt Singapore when the financial crisis hit. However, in June 2010, the increase in global demand raised the country's exports to Europe by 75%, to China by 39%, and to Japan by 50% when compared to the previous year (Shannon 2010). In general, the 2008 global financial crisis has not significantly impacted Singapore's medium-term growth performance. Its economy rebounded strongly in 2010, primarily driven by external demand, as Singapore's non-oil exports increased by 24% in 2010 (Asher 2010). Singapore's manufacturing sector also expanded 46% in the second quarter of 2010, which was led by biomedical production gains (Shannon 2010). The goods-producing sectors, including the largest contributor—electronics manufacturing—have been the ones credited with such a fast upturn. Overall, an increase in exports, tourism, and a booming manufacturing sector was the main drivers of Singapore's economic expansion (Shannon 2010). In 2010, Singapore was the third fastest growing economy in the world behind Qatar and Paraguay (Economy Watch 2010b), with a real GDP growth rate of 12.46% (Fig. 2.1), unemployment rate of only 2.2% (Fig. 2.3), and consumer price index inflation of 2.8% (Fig. 2.4).

In the third quarter of 2011, the Singapore economy grew by 6.1% on a year-on-year basis up from 1.0% in the preceding quarter (MTI 2011). However, the pace of expansion slowed significantly in 2011 as the global economy faced difficulties due to the sovereign debt crisis in some European economies and a slow pace of growth in the United States (Venkat 2011). Singapore's Ministry of Trade and Industry predicted that although resilient domestic demand in emerging Asian countries would provide some support to global demand, it would not fully mitigate the effects of an economic slowdown in the advanced economies (MTI 2011). Consequently, Singapore's externally oriented sectors such as electronics and wholesale trade would continue to slow, while the financial services sector would be affected by heightened uncertainties in the external environment. Nevertheless, in 2011 Singapore was ranked third in the global competitiveness behind Switzerland and Sweden (Schwab 2011).

Moving forward, external demand is expected to be less robust due to lower anticipated growth in the U.S., Europe Union, and Japan. However, Singapore is expected to partially offset this with the projected high growth rates in ASEAN countries (including Brunei, Burma, Cambodia, Indonesia, Laos, Malaysia, the Philippines, Singapore, Thailand, and Vietnam), China, and India. Internally, Singapore's growth would rely on the biomedical manufacturing cluster and tourism-related sectors (MTI 2011). In the near future, it is likely that Singapore will see accelerated inflation in sectors such as food and housing, tightening of monetary policy, and further appreciation of the Singaporean dollar. To cope with these challenges, Singapore is already attempting to diversify its economy and establish several niche areas, including education, wealth management, and entertainment (Asher 2010). Another thrust is Singapore's planned infrastructure spending, which includes around US$46 billion (S$60 billion) over the next decade in expanding transportation networks (Asher 2010). This endeavor not only helps the country wade through the downturned economy but also shows its ambition to strengthen Singapore's position as a global city.

Over the last decade, Singapore has attracted major investments in pharmaceuticals and medical technology production and will continue efforts to establish itself as Southeast Asia's financial and high-tech hub. For the longer term, the Singaporean government hopes to establish a new growth path that focuses on raising productivity, which has sunk to 1.8% growth per year in the last decade (CIA 2012).

The above planning and flexibility may assist Singapore in continuing its growth and resilience. As Professor Thangavelu (2009) at National Singapore University put it, "In a long and protracted recession, the economy needs to restructure and new industries with new technologies must emerge to create a robust economic upturn." Thangavelu (2009) also predicted that the 2008 financial crisis would provide an opportunity to develop stronger and more innovative SMEs in Singapore.

Taiwan

In Taiwan, signs of recovery from the financial crisis started to show by the end of 2009. According to the Central Bank of Taiwan, Taiwan's economy experienced a strong recovery in 2010, based largely on growing Asian-led demand and tightened commercial links with China (Cooke 2012; Global Finance 2011). As the global economy was recovering, the demand for downstream products was picking up and the manufacturing sector was expected to see a slight boost in sales. In 2010, Taiwan's exports were projected to grow 10.51% and imports by 10.13%, which would translate into a trade surplus of US$32.74 billion (Cheng and Kao 2010). The Taiwan Institute of Economic Research also projected that with consumer confidence climbing and improved unemployment, private consumption was expected to grow 2.07% in 2010 (Cheng and Kao 2010).

In the spring of 2010, the Asian Development Bank raised its growth forecast for Taiwan to 4.9% for 2010, compared with the 3.5% that it predicted in December 2009. To further stimulate the economy, Taiwan's government reached a consensus over a draft proposal to reduce corporate tax from 20% to 17% in April 2010. In reflection, a strengthening global economy coupled with Taiwan's ability to deliver new electronic products to the world bode well for export growth. In June 2010, Taiwan and China signed the Economic Cooperation Framework Agreement (ECFA). This is an agreement that Taiwan authorities hope will lead to free-trade arrangements that will increase economic ties with China by lowering tariffs on a number of goods and by reducing market access barriers for services. Closer economic links with China bring greater opportunities for Taiwan's economy, but, at the same time, it also poses new challenges. For instance, Taiwan's increasing investment in China has resulted in Chinese import substitution away from Taiwan's exports and a restriction of potential job creation in Taiwan (CIA 2012).

By November 2010, Taiwan's unemployment rate had fallen to a 2-year low of 4.73% and industrial output was up 19.37% from a year earlier. IMD database reported that Taiwan's 2010 real GDP growth per capita was 10.62%, as exports returned to the level of previous years. In 2010, Taiwan's total trade surplus was US$23 billion (Global Finance 2011).

In 2011, the improving global macroenvironment and the boost from signing the ECFA (Economic Cooperation Framework Agreement) with China, Taiwan's export growth continued. According to the Director General of Central Bank of Taiwan, domestic demand, consumers confidence, and spending have been increased with the economic recovery (Global Finance 2011). After the financial crisis, Asia as a whole bounced back a lot quicker than the West. Taiwan has performed particularly well in trade exports and tourism. Boosted by the strengthening economic ties between Taiwan and China, Taiwan's tourism has become the second biggest growth engine in its economy (Global Finance 2011). It is now a creditor economy, holding the world's fourth largest stock of foreign exchange reserves—US$399.5 billion as of April 2011 (Global Edge 2012).

According to the Global Competitiveness Index published by World Economic Forum, Taiwan remains unchanged at no. 13 in both 2010–2011 and 2011–2012 reports. Taiwan's economy continues to be supported by strong business innovation and improving financial market developments (Schwab 2011). Regionally, Taiwan ranked fourth behind Singapore, Japan, and Hong Kong, while outperforming South Korea. In the future, Taiwan's continued development depends on further transformation to a high-tech and service-oriented economy as well as continued diversification of its trade markets.

China and Hong Kong are Taiwan's largest export markets (refer to Appendix 8), followed by the United States, and then Japan. Three financial memorandums of understanding with China, covering banking, securities, and insurance, took effect in mid-January 2010. This step opened Taiwan to greater investments from China's financial firms and institutional investors while providing new opportunities for Taiwanese financial firms to operate in China. Despite the continued economic slowdown in Europe, Taiwan's government projected that it should be free of the so-called double-dip recession and would remain an Asian economic frontrunner no matter how the global economic situation unfolds (Liu and Wu 2011).

Chapter 5
Future Perspectives and Policy Implications

By the middle of 2012, more than 3 years have passed since the outbreak of the 2008 global financial crisis. Although some countries in the European Union are still in deep financial trouble, it is time to reflect upon what happened during the past few years and what can be learned from this crisis for future preventive actions.

In January 2012, the International Monetary Fund announced that the global recovery was threatened by intensifying strains in the Euro area; thus, financial conditions have deteriorated, growth prospects have dimmed, and downside risks have escalated (IMF 2012). The Euro area economy was expected to go into a mild recession in 2012. The effects of bank deleveraging on the real economy may surface again, and the impact of additional fiscal consolidation may heighten.

Even growth in developing economies that had traditionally remained resilient was expected to slow down because of the worsening external environment and a weakening of internal demand (IMF 2012). The global investment prospects continue to be clouded by the significant uncertainties with the sliding economy. In contrast to this, China has emerged stronger from the global financial crisis and its continuous internal markets growth has increased trading within the four Greater China economies. In addition, the economic crisis has provided an ideal opportunity to accelerate the integration of the ASEAN Economic Community in Asia (Thangavelu 2009).

In this chapter, we first present the prospects of these four economies and then describe their potential challenges in the future. The prospects and challenges are meant to provide points of thought for discussion and do not represent an exhaustive list. This chapter finishes with some policy implications.

C.Y.-Y. Lin et al., *National Intellectual Capital and the Financial Crisis in China, Hong Kong, Singapore, and Taiwan*, SpringerBriefs in Economics 8,
DOI 10.1007/978-1-4614-5984-2_5, © The Author(s) 2013

Prospects

In 2010, Asia is already the second largest center of world trade, and by 2020, it will likely be the largest (Drysdale 2010). Intra-regional trades in ASEAN and Asia are key factors in export growth as export markets in the United States and European Union lag. According to the World Trade Organization (WTO), intra-regional merchandise trade in Asia accounts for almost half (49.7%) of Asia's total exports in 2007, while Asia's exports to North America and Europe account for 19.9% and 18.8%, respectively (Xu 2010). In addition, trade within Asia has experienced a significant shift from traditional interindustry trade to intra-industry trade in finished goods and, more recently, to intra-industry trade in parts and components (Xu 2010). This situation is particularly true within the Greater China economies. For example, annual trade between China and Taiwan has increased from less than 3% in 2000 to 26% in 2008 or 40% if including Hong Kong (Xu 2010).

China

During the financial crisis, China's strong recovery has pulled the world economy along. Ten years ago, China's foreign exchange reserves totaled US$166 billion, or 14% of its GDP; in 2010, its foreign exchange reserves towered at US$2.85 trillion, or half of its GDP (Herd et al. 2011). OECD reports that China's contribution to global growth has been increasing exponentially, from 5% in 1980 to 12% in 2000 and 30% in 2010. Almost 1.5 percentage points of the projected growth of 4–4.5% of the world economy in 2011–2012 is accounted for by China (Herd et al. 2011). With strong financial power, China's prospects for the years to come may include the points listed in the following paragraphs. These points are mainly based on OECD research (Herd et al. 2011) and Heilmann and Schmidt's report (2010).

First, China's response to the crisis contains a broader concern for development issues. 9% of the stimulus package allocated for technology advancement and restructuring would assist China to use the opportunity to push forward and become a stronger country after the crisis.

Second, China's ambitious 12th Five-Year Plan, running from 2011 to 2015, seeks to transform the economy from a labor-intensive and investment-led powerhouse focused exclusively on GDP growth to that based on more dynamic and technology-based industries. Its target is to build a more sustainable model that balances growth with social harmony and innovation with environmental protection. The resources devoted to science and technology in China have expanded rapidly in recent years; it now ranks among the top countries in total R&D spending, as well as total number of researchers.

Third, China also puts great emphasis on generating more equitable growth across the different regions, between rural and urban areas and within urban areas. It acknowledges that increasing domestic demand and improving the income distribution will require a comprehensive development of social services.

Fourth, although China still lags in human capital among the studied 48 countries, it has invested heavily in education. For example, one of China's biggest success stories has been the implementation of universal 9-year education to age 15; this was achieved in 2009 during the financial crisis. Consequently, there has been a massive change in the quality of the labor force within China. Moreover, almost three-quarters of children attend senior secondary schools to the age of 18. Furthermore according to China's plan, a determined effort is committed to upgrade the quality of education in the poorer parts of the country.

Fifth, China is shaping its competitive environment in multiple ways. To ensure that state-owned enterprises and their suppliers (in industries such as construction, equipment, and steel) have secure revenue streams, China's central government planning includes laying out clear plans for expanding the penetration and capacity of China's rail, power, and communications networks. Its Ministry of Finance is prepared to be a minority shareholder in venture capital funds if the fund focuses on start-ups in designated sectors from environment protection to new-energy vehicles (Reuters 2011). Such planned infrastructure improvement assists its urban as well as rural development. Other expansion covers industries such as education, financial services, health care, and logistics. Relevant policies are being deregulated, which opens the market to foreign companies. In addition, increased attention to food, drug safety, and quality guides companies to pursue high-quality standards.

Sixth, the Chinese government realizes that creative industries provide China with a new perspective on how to narrow the gap vis-à-vis developed countries and a new way to explore the potential of development as a big country rich in cultural resources. As a result, the Chinese government gives money to some venture capital investment firms to boost the development of creative industries (Reuters 2011).

Seventh, China has strong outward foreign direct investment. These investments have soared in the past 5 years and even rose very slightly in 2009 when other countries' FDI nearly halved during the financial crisis.

It is clear from this financial crisis that China's economic cycle is increasingly influenced by, and induces, fluctuations in the rest of the world. To progress together with China's future prospects, other countries may probe the implications of China's 12th Five-Year Plan for their own industries to identify the opportunities and risks ahead.

Hong Kong

The prospect of Hong Kong is closely connected with the economic development of China as follows:

First, Hong Kong is still far ahead of all other major cities in China including Shanghai in terms of global accessibility and international competitiveness. As a result, Hong Kong plays an important role by serving as a gateway to China as well as linking China with the rest of the world. It will further strengthen its role as a dominant service-regional metropolis and become the top international financial center for China. This not only includes the banking/financial services provisions but, more importantly, includes the mobilizations of China's inward and outward international investments with the global world (Tuan and Ng 2011).

Second, the banking sector plays a prominent role in Hong Kong's economy. While the sector accounts for only 2.5% of employment, its share of nominal GDP has been rising steadily from 7.5% in 1997 to almost 12% in 2007. Hong Kong's banks experienced strong growth throughout recent years as they expanded their business to mainland China.

Third, China allows local Hong Kong banks to conduct more business in RMB for corporate clients. In addition, local business firms can issue RMB corporate bonds and apply for bank loans and credit lines in RMB to finance projects as well as to settle trade in RMB. These moves definitely facilitate Hong Kong as a testing ground for the internationalization and convertibility of the RMB currency. Furthermore, it will enhance the development of Hong Kong's critical role as an offshore RMB center (Tuan and Ng 2011).

Fourth, the growing China mainland market will stimulate commodity exports and services from Hong Kong. The gradual opening of Chinese markets will provide increasing opportunities for Hong Kong investors. Thus, combining China's economic power with Hong Kong's established infrastructure, products/instruments, well-trained manpower, global accessibility, and business networking, their synergy will be significant.

Fifth, continuous improvement in bilateral trading environments with other Asian countries, such as ASEAN, will also benefit Hong Kong in both the external trade of commodities and services in the long run.

Sixth, according to the 2012 index of Economic Freedom, Hong Kong is the freest economy in the world (Heritage Foundation 2012).

Singapore

Singapore is a city nation, small in area (697 square kilometers only) with a relatively small domestic market (around 5.35 million people, July 2012). Yet, it is both high in GDP and national intellectual capital with the following prospects:

First, Singapore has capitalized on its geostrategic position to become the world's transport center. It is the busiest port in the world, above Rotterdam and Hong Kong (Economy Watch 2010b).

Second, many multinational companies have set up their regional headquarters in Singapore due to its high degree of globalization and accessibility to other markets.

Third, with an inherent vulnerability to external impact, Singapore has an efficient government to enact economic policies that safeguard the country from perturbations in the global market.

Fourth, with a small domestic market, Singapore leverages ASEAN as its hinterland with a population of nearly 540 million (about 100 times of its own). This strategy has helped Singapore recover faster than expected from this global financial crisis. The Singaporean economy has demonstrated astounding resilience to the 1997 Asian financial crisis and this 2008 global financial crisis.

Fifth, Singapore is the only Asian country to have AAA credit ratings from all three major credit rating agencies—Standard & Poor's, Moody's, and Fitch (Economy Watch 2010b). With such a reputation, its role as a financial service hub can be assured.

Sixth, Singapore has the highest trade to GDP ratio in the world at 407.9% (Economy Watch 2010b). According to the 2012 index of Economic Freedom, Singapore is the 2nd freest economy in the world (Heritage Foundation 2012).

Taiwan

With very limited resources, Taiwan recovered from the 1997 Asian financial crisis and the 2008 global financial crisis relatively fast. Its prospects may include the following:

First, Taiwan was ranked third-best globally for its investment environment in a 2011 report by Business Environment Risk Intelligence.

Second, Taiwan is strong in innovation. According to World Economic Forum 2011 report, Taiwan placed first in utility patents granted for the fourth straight year (Schwab 2011).

Third, Taiwan has a strong semiconductor industry and overtook that of the United States, second only to Japan in 2007. In 2009, Taiwan ranked first in global market share for IC manufacturing, packaging, and testing, as well as second place in IC design. Due to its strong capabilities in OEM wafer manufacturing and a complete industry supply chain, Taiwan has been able to distinguish itself from its competitors.

Fourth, Taiwan runs a large trade surplus, and its foreign reserves are the world's fourth largest, behind China, Japan, and Russia (CIA 2012).

Fifth, Taiwan's Ministry of Economic Affairs offers a variety of tax incentives aimed at encouraging corporate investment in research and development, personnel training, and new equipment and technology to upgrade industrial technology.

Sixth, 98% of all companies in Taiwan are considered SMEs (small and medium enterprises) with strong entrepreneurship, which facilitates its developing into an innovation-driven economy.

Challenges

Alongside the prospects, there are challenges as well in the Greater China region as briefly described hereunder.

China

After recovering from the financial crisis, China faces some immediate economic challenges as follows:

First, during the financial crisis, China's central government allocated heavy stimulus share to the local governments, some of them became so over-indebted that the central government might be forced to pick up the bill.

Second, China's lending system favors primarily the big state-owned enterprises. Cash-strapped private SMEs continue to face poor borrowing conditions, leaving them with little choice other than to turn to the informal financial market (Heilmann and Schmidt 2010).

Third, inflation has long been a problem for China. The largely fixed exchange rate system makes counterinflationary monetary policies less flexible. In October 2011, Chinese leaders announced that reining in inflation remained an overall priority, dampening expectations that policymakers will attempt to boost growth by loosening credit, lowering interest rates, or lifting restrictions in China's residential property market (Pierson 2011).

Fourth, talent shortage is another concern. By 2025, China is expected to have more than triple the number of university graduates it produces every year, many of these degree holders will lack the skills that top Chinese companies and multinationals demand (Devan et al. 2008). With rapid economic development, smaller cities will face intense pressure to attract and create appropriate jobs for an educated workforce. Attracting qualified workers to smaller cities helps diversify the economy and shield the smaller cities from future decay that the U.S. Rust Belt experienced during the late twentieth century. However, it poses a great challenge.

Fifth, aging in China increasingly has become an issue, driven largely by low fertility rates of the one-child policy. Among the four BRIC countries, population percentage under 14 years of age and median age are 26.2%/29.3, 15.2%/38.7, 29.7%/26.2, and 17.6%/35.5 for Brazil, Russia, India, and China, respectively (CIA 2012). That is, both Brazil and India have more population under the age of 14 with median age under 30. Although China's population structure is a little better than Russia, the aging problem entails not only decreasing workforce but also an increasing financial burden for the government.

Sixth, inequality and urbanization also is a point of growing concern. For the next 10 years, urbanization will be a striking dimension of China's rapid ongoing transformation. Continued urbanization will boost incomes, consumption, and the size of the service sector, thereby increasing the domestic market and helping with external imbalances. But it calls for further labor market and social policy reforms to facilitate internal migration and reduce inequality. Some of these, notably in the areas of pensions, health care, and the provision of education for migrants' families, need a sizeable financial support. Another inequality is that rural China remains poor even as its cities increase in affluence. Accompanying China's rapid economic transformation, inequality has aroused social strife to a certain degree.

Seventh, the environment is one of China's three most serious challenges in addition to the transition to a truly globalized market economy and political reform (Fleet 2010). Environmental damage—notably air pollution, soil erosion, and the steady fall of the water table—is watched by environmentalists. Other than industrialization, urbanization is a key factor of the environmental issue. Harmonious urbanization also raises other challenges, not least on the environmental front. Making the right choices in this respect should help China reach its energy intensity reduction targets (Herd et al. 2011).

Eighth, the Chinese government acknowledges that corruption is endangering political stability and economic growth and has been pursuing an anticorruption campaign to curb this persistent problem. According to the Business Anti-Corruption Portal (2012), China's transition from planned to market economy not only led to the unprecedented economic growth but also created opportunities for public officials and others to enrich themselves illegally.

Hong Kong

Hong Kong has recovered well from this global financial crisis through being a part of China. Yet, on the eighth meeting of the Commission on Strategic Development, the issues of housing, inflation, aging society, and the career development of young people were particularly emphasized (Lau 2011). In addition, Hong Kong faces the following challenges as well:

First, economic marginalization is a future challenge for Hong Kong. A former governor of the Executive Council of Hong Kong is worried that Hong Kong's relative superiority is descending, mainly for the reasons that Hong Kong is no longer a free land market being supervised by the Chinese government (Leung 2008). Furthermore, China plans for Shanghai to catch up to New York as a global financial center by 2040–2050. This poses another threat to Hong Kong's marginalization. Therefore, Hong Kong needs to consider how to fully display its existing advantages to further strengthen its status as a financial, trade, and shipping center.

Second, Hong Kong needs to fully make use of its geographical advantage to strengthen its ties with the Pearl River Delta and the mainland China at large.

Third, Hong Kong needs to develop its competitive industries, based on its own strength, such as its service sector.

Fourth, Hong Kong's aging problem is even more serious than that of China's, as its population under the age of 14 is only 11.6% and the median age is 43.4 (CIA 2012).

Singapore

For Singapore, Prime Minister Lee Hsien Loong predicted that a 5% growth rate over the next decade would be difficult to sustain (Barta and Holmes 2010).

Singapore's leaders have become more cautious about the outlook in the aftermath of the financial crisis. There is special concern that Singapore's export-dominant economy leaves it too reliant on swings in consumer demand elsewhere. The following challenges may pose some threats as well:

First, Singapore's reducing productivity is a great concern. Many Singaporean employers count on cheap foreign workers to keep their costs low. Yet with a lower quality labor force, labor productivity growth fell from 3.1% annually during the 1990s to 1.3% annually during the 2000–2009 period. Consequently, Singapore economic strategy committee advised to raise levies on foreign workers to gradually force employers to invest in more productive operations, with a goal of achieving 2% to 3% annual productivity growth and to call for more investment in research and development (Barta and Holmes 2010).

Second, in the coming decade, Singapore's population will age rapidly, with 13.8% of the population under the age of 14 and a median age 40.1 in 2012 (CIA 2012). Its total fertility rate has stayed between 1.2 and 1.3, with zero net in immigration; this means that population will halve in 45 years. The low fertility and aging issue is also connected to Singapore's reliance on foreign manpower to sustain growth. Predictably, the share of foreign workers within the total labor force has already exceeded one-third and, at current rates, could exceed 50% in coming decade (Asher 2010).

Third, with an aging population, low fertility rate, and lengthened life expectance (due better health care), the workers may need to extend their retirement (currently 62 years of age). With a lengthened retirement age comes the challenge to motivate and train older workers.

Fourth, with a religious profile of 42.5% Buddhist, 14.9% Muslim, 9.8% Christian, 8.5% Taoist, 4.8% Catholic, and 4% Hindu (CIA 2012), Singapore has to guard against forming religious enclaves and mix with those of other faiths to prevent society from becoming divided (Goh 2009).

Taiwan

Taiwan had a quick turnaround after the 2008 financial crisis. Yet, the following challenges may pose some threats for its future development:

First, the "Hollowing of Industry" is a serious industrial and economic problem in Taiwan. Due to its past successful economic development, some developing countries are offering very large tax incentives to attract many cheap-labor mass production OEM industries and businesses away from Taiwan. With the increasing labor cost, the majority of the cheap-labor industries have moved out, forcing the industries that are still in Taiwan to transform to more intellectual/knowledge intensive industries.

Second, developing intellectual/knowledge intensive industries also poses a great challenge to Taiwan, as it will have head-on competition with the knowledge-based advanced countries. Although Taiwan has already established a strong foundation for its electronic and information industries, it still suffers from a shortage of high-technology manpower.

Third, Taiwan needs to withstand commoditization pressures of electronic and information products in the fast-moving IT industry. Taiwanese manufacturers will need to stay close to consumers in China and other key markets to maintain their recent pace of innovation in mobile handsets and notebook computers (Cooke 2012). Furthermore, Taiwan also faces competition from South Korea and Singapore in its reindustrialization effort.

Fourth, Taiwan's total market value of foreign holdings in stocks, government papers, and bank deposits stands at US$254 billion or the equivalent of 66% of Taiwan's foreign exchange reserves at the end of 2010 (US$382 billion), according to the Director General of the Central Bank of Taiwan (Global Finance 2011). This high percentage of foreign holdings makes Taiwan very vulnerable to the international financial market fluctuation.

Fifth, over the past decade, a relatively high proportion of Taiwan's capital has gone offshore and stayed offshore. In addition, there is a lot of money generated and produced in Taiwan that actually ends up in the financial centers of Hong Kong and Singapore. So, while Taiwan has managed capital flows very well, these trends might not have been effective in maximizing GDP and GNP growth (Global Finance 2011).

Sixth, reliance on a narrow range of trading partners (mainly China, the U.S., Japan, and the West) has made Taiwan highly vulnerable to their demands. As of the end of April 2011, Taiwan firms had invested about US$102.1 billion in China, which is more than 60% of Taiwan's stock of direct foreign investment (Global Edge 2012). In addition, the U.S. calls for import liberalization and currency revaluation have hurt the export-dependent Taiwanese firms. Heavy reliance on exports also makes Taiwan extremely vulnerable to downturns in the world economy, as was the case with the 2008 global financial crisis.

Seventh, Taiwan's population is aging quickly as well, with the number of people under 14 accounting for 15.6% and median age of 37.6 in 2011. Taiwan's total fertility rate of just over one child (1.15) per woman is among the lowest in the world, raising the prospect of future labor shortages, falling domestic demand, and declining tax revenues.

Policy Implications

In the wake of the 2008 global financial crisis, many potential causes for it have been suggested. The United States Senate issued the Levin-Coburn report (Levin and Coburn 2011) stating "The crisis was not a natural disaster, but the result of high risk, complex financial products; undisclosed conflicts of interest; and the failure of regulators, the credit rating agencies, and the market itself to rein in the excesses of Wall Street." Critics argued that credit rating agencies and investors failed to accurately price the risk involved with mortgage-related financial products and that governments did not adjust their regulatory practices to address twenty-first-century financial markets.

The above statements manifest that intangibles—people, systems, governance, market dynamics, and the global interdependence—are the keys that caused this global financial catastrophe. National intellectual capital (NIC) represents national intangible assets to a certain degree. Research results reported in this booklet provide valuable information for policymakers to look back on the national intangibles before the crisis, observe their codevelopment with the GDP during and after the crisis, and then draw lessons from it.

By investigating economic development and NIC development from 2005 to 2010, we have been able to connect the coevolution of tangibles and intangibles for the four Greater China economies. Implications that can be drawn from our research findings are as follows:

First, the Greater China economies are on solid ground progressing toward their continuous economic recovery and future development. The objective statistics (Figs. 2.1, 2.2, 2.3, and 2.4) show the impact of the global financial crisis on these four economies. That is, real GDP growth was down, general government debt was up, unemployment rate was up, and consumer price inflation was up. However, their economies all rebounded swiftly. For China and Hong Kong particularly, their general government debt after 2008 was even lower than its 2007 levels, and China's unemployment rate was not affected too much (around 4.0–4.3% through the 6 years). As a result, these four economies are on a relatively solid ground progressing toward their economic recovery and future development. In particular, China and Hong Kong will not be burdened by the public debt that impedes financial crisis recovery concluded in Reinhart and Rogoff's (2009) research on financial crises over the last 800 years.

Second, the Greater China economies have become a powerful integrated economic entity that can sustain itself. From the reported statistics in 2009 and 2010, the four economies showed very clear signs of speedy recoveries. The financial stimulus plans seem to work in the Greater China region with immediate results. This is in spite of the fact that Hong Kong, Singapore, and Taiwan were all hard hit by the financial crisis due to their heavy reliance on export and the sudden, drastic export demand reduction. Their successful recovery was partially due to the gradual recovery worldwide, yet the main driver was their close ties with China in which all three countries directly benefited from the resilience and the speedy recovery of China. This is particularly true of Hong Kong, as it is already a part of China. China's huge and continuously growing domestic market has provided the other three economies the best export outlets. Readers can refer to Appendix 8, which summarizes the percentage of export and import partners within the four economies, for how heavily the three small economies rely on China.

Third, the Greater China economies can collaborate in NIC development to reap a synergic effect for their next stage economic development. According to the World Economic Forum, Hong Kong, Singapore, and Taiwan are all in Stage 3 development, i.e., innovation-driven economy. Therefore, future competition is based more on intangibles (e.g., human capital, R&D, and innovation) than tangibles (e.g., machines, facilities, and land). With data from 48 countries, we find that the higher the NIC, the higher the GDP. This clearly manifests that intangible NIC advancement accompanies GDP growth and should be considered as a national strategic issue. For these four economies, each capital in each economy has different types of progression. For example, renewal capital in Singapore correlates better with GDP growth than it does in Taiwan, even though its renewal capital is lower than that of Taiwan's. In terms of the added value of renewal capital, Taiwan needs to learn from Singapore. Singapore, in turn, can advance its innovation and entrepreneurship from studying Taiwan's experience. Our research results provide directions for complementary effects among the four economies and for them to develop their NIC in a more focused and efficient way.

Fourth, Hong Kong, Singapore, and Taiwan can benchmark the Nordic region for long-term oriented human capital and renewal capital development. Hong Kong, Singapore, and Taiwan are categorized as innovation-driven economies, according to World Economic Forum. To develop knowledge-based and innovation-powered economies, the long-term oriented NIC is of prime importance. However, Fig. 3.14 shows that for human capital and renewal capital codevelopment, Singapore and Taiwan are roughly in the second quarter, Hong Kong is in the third quarter, and China is in the bottom quarter of the 48-country continuum. There is still room for improvement, even for Singapore and Taiwan. The top quarter Nordic countries are the best examples for the three small economies to benchmark for their next stage of NIC development.

Fifth, NIC international competitiveness can be regularly monitored for strategizing national development and well-being. Internal NIC progression or regression is not necessarily reflected in NIC international competitiveness. Taking China as an example, a small 0.5 point score decline in its market capital leads to the loss of 10 ranks and a 1.0 point score reduction in process capital results in the loss of 9 ranks, as shown in Fig. 3.17. Tangibly, China is resilient and has strong economic power as evidenced during this global financial crisis; however, China lost ground in the global NIC standing during the financial crisis. On the contrary, Singapore being hit the hardest in the financial crisis still stood relatively firm in its market capital and process capital international ranking and GDP growth. Apparently, both solid tangible and intangible foundations may help withstand economic downturn and continuously contribute to national development and well-being, as is the case of Singapore. Therefore, a country's NIC international competitiveness can be regularly monitored for timely coping strategies.

Sixth, NIC mapping and identification of the key efficiency drivers provide directions for linking NIC development to the bottom line GDP growth. For developing knowledge-based or NIC-based competitiveness, the advancement of a nation's NIC needs to accompany with its bottom line GDP growth. However, some economies failed to show such ideal codevelopment. For example, Taiwan's renewal capital progressed faster than its GDP growth. That is, its renewal capital improvement does not accompany with GDP growth. In addition, the 3D trajectory analysis discloses efficiency drivers for each economy (Table 3.4). For example, for China's NIC to facilitate its GDP growth, the fundamentals such as computers in use per capita and higher education enrollment should be emphasized more; for Hong Kong, more emphasis should be placed on R&D; and for Taiwan, more efforts need to be put on government-related improvement, such as government efficiency and policy transparency. The NIC mapping and the efficiency drivers presented in this booklet provide clear directions for a more focused national development policy.

Seventh, leveraging NIC to secure national well-being is a winning strategy for resource-strained economies. Singapore provides the best example of smartly utilizing its NIC to obtain a unique competitiveness. Strictly speaking, Singapore's only tangible asset is its deepwater port. Most of its daily necessities need to be imported, including water. Without any oil fields, Singapore has the third largest oil refinery in the world, behind Rotterdam and Houston only (Economy Watch 2010b). Its smart people (human capital) leverage Singapore's geostrategic position to achieve the number one market capital and create its current high GDP per capita (ppp), next to only Norway. During this financial crisis, Singapore successfully leveraged ASEAN countries as its hinterland to swiftly rebound from the financial shock. These are two examples of how Singapore sustains its competitiveness through leveraging its intangibles, which may shed some light for resource-constrained economies.

Eighth, the Greater China economies can capitalize on their synergy to form a value chain and provide a complete spectrum of services to the rest of the world. From this global financial crisis study, it becomes more evident that a nation cannot fight alone. Having partners and allies can form a self-sufficient and semidomestic market in which member countries can support each other. Therefore, to build a more resilient economy, consideration can be placed on identifying critical partners for benchmarking, cocreating, and forming a better functioning economic region. To demonstrate how the four Greater China economies can supplement each other's resources, Fig. 5.1 shows a much longer 16-year (1995–2010) NIC progression. The connected line of the three small economies presents the value chain concept. As a team, they can supply a spectrum of services that required by countries at different stages of development. With the speed of China's economic development, China should be able to fill the line gap in between relatively fast. At that time, the four economies will have a complete spectrum of value chain and experience, solidified by their Chinese origin (95%, 76.8%, and 98% ethnic Chinese in Hong Kong, Singapore, and Taiwan, respectively). Its synergic economic power in the future global economy should be enormous.

Concluding Remarks and Emerging Insights

In retrospect, according to U.S. federal inquiry, the 2008 financial crisis was an "avoidable" disaster caused by widespread failures in government regulation, corporate mismanagement, and heedless risk-taking by Wall Street (Chan 2011).

Fortunately, a great recession was avoided due to national and international measures that prevented the financial crisis from deteriorating. The Greater China

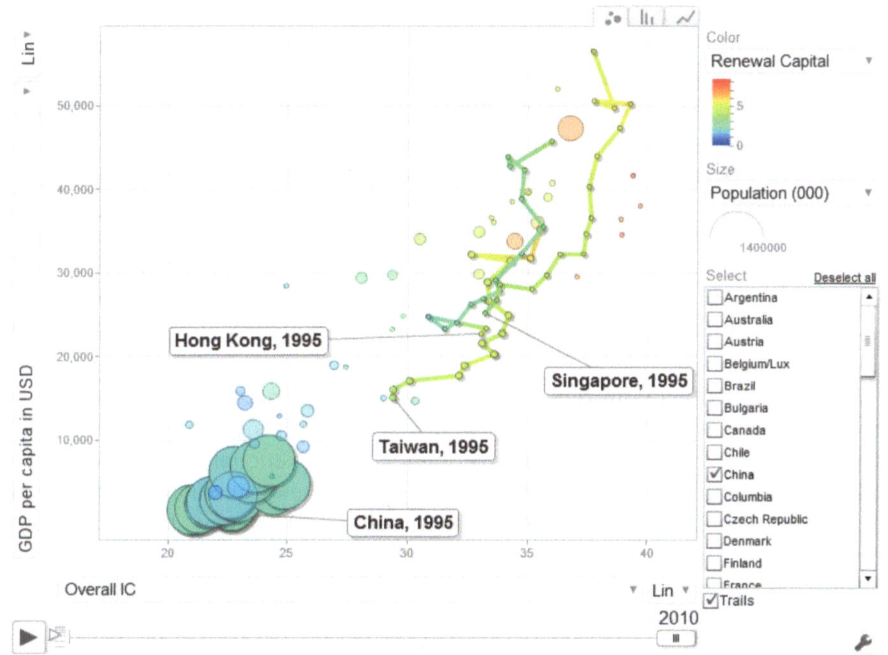

Fig. 5.1 The synergic effect of the four Greater China economies

economies have proven to be relatively resilient in comparison to countries in other regions. In the aftermath of this global financial crisis, there is a worldwide expectation that China should play an increasing role in stabilizing the global economy, not only for its financial support but also for its growing internal markets and for its outflow of investments to rescue some dying industries or regions in various countries. The very origin of the financial crisis—people, systems, governance, market dynamics, and the global interdependence—is also the key to the economic recovery in any place and country. For China to successfully perform its expected role at this historical moment, its capabilities augmented by other Chinese ethnic economies may facilitate Asian and global economic development.

As a recap, to counter worsening economic conditions worldwide, a total of US$2 trillion in stimulus packages, amounting to approximately 3% of the world gross domestic product, has been allocated (some of the packages include measures that extend into subsequent years), an amount that exceeds the call by the International Monetary Fund (IMF) for fiscal stimulus totaling 2% of global GDP. IMF also estimated that as of January 2009, the U.S. fiscal stimulus packages as a percentage of the GDP in 2009 would amount to 1.9%, for the Euro area 0.9%, for Japan 1.4%, for Asia excluding Japan 1.5%, and for the rest of the G-20 countries 1.1% (Nanto 2009).

In terms of future NIC development, our other study of the top ten NIC ranking countries reveals that the interplay among the four capitals—human capital, market capital, process capital, and renewal capital—resulted in the best synergic prospect for GDP growth. When their weaker capital (process capital) interacts with stronger capital (market capital), both effects on the GDP are enhanced (Lin 2011). This finding prompts our suggestion for collaborating NIC development among these four economies.

It is a fundamental function of government to create and maintain both tangible and intangible infrastructure under which the economic, social, and cultural life of the citizens can prosper. Looking back at the 2008 global financial crisis through the lens of national intellectual capital provides a different perspective for the policymakers to ponder on.

At the completion of this reporting, we have a strong understanding that national success is a capability that can be accumulated meticulously over time; it is not a free gift nor can it be achieved from overnight wealth. National economic development is built upon the well-coordinated intangible systems, including human capital, market capital, process capital, and renewal capital systems described in this booklet. Since it takes time to develop intangibles, countries that are determined to enhance their so-called soft-power need to consider the added value issue from the start of their policymaking.

Appendices

C.Y.-Y. Lin et al., *National Intellectual Capital and the Financial Crisis in China, Hong Kong, Singapore, and Taiwan*, SpringerBriefs in Economics 8,
DOI 10.1007/978-1-4614-5984-2, © The Author(s) 2013

Appendix 1 Summary of the main stimulus packages of the Greater China economies

Country	Date	Amount	Item	Sponsor and remarks
China	Nov 9, 2008	#1 #2 #3	Stimulus package—plan as of March 2009:	More than 86% of the PRC's stimulus package went to infrastructure spending, out of which 45% was for road, rail, and airport infrastructure (US$ 263.7bn) (CNY1.8tn); 9.5% was for improving electricity, water, and road infrastructure in rural areas (US$54.2bn) (CNY370bn); 7% was for low income housing (US$41bn) (CNY280bn); and 24.7% was for the reconstruction of towns devastated by the May 12, 2008 earthquake. The remainder of the stimulus package went to health care and education (US$5.9bn) (CNY40bn or 1% of the total package), ecological and environmental protection (US$51.3bn) (CNY350bn or 8.8%), and technical innovation (US$23.4bn) (CNY160bn or 4%)
		US$586bn (CNY4tn) (19% of 2008 GDP)	25% Sichuan earthquake reconstruction; 38% public infrastructure, 10% housing for low income; 9% rural development; 9% technological innovation and structural adjustment; 5% sustainable development (energy, environment); 4% health care, education, culture, family planning	* CNY is the formal name of Chinese currency, also known as RMB
	Jan 12, 2009	#1 US$87.8bn investment spending		Scientific and technical innovation/upgrades
	Jan 21, 2009	#1 US$124.3bn		Health care reform to provide basic medical security for all citizens, improve the quality of medical services, and make medical services more accessible

Hong Kong	Dec 2008	#4 US$12.9bn	Stimulus package	For small and medium-sized enterprises. The business loan ceiling was raised to US$769,230, while more than 60,000 jobs would be created in 2009
	2009–2010	#5 US$20.6mn (HKD1.6bn) US$108.4mn (HKD840mn)	Stimulus package	The government earmarked US$20.6mn (HKD1.6bn) for the creation of 62,000 jobs, taxpayers in Hong Kong received a one-off rebate of US$774 (HKD6000). Other measures include a 20% cut in the cost of renting property from the government, an increase in expenditure of medical services of US$108.4mn (HKD840mn)
Singapore	Jan 2009	#1 US$13.7bn	Stimulus package	25% of Singapore's stimulus package was spent on infrastructure, 12.5% on enhancing future capacity, 25% was spent as loan guarantees to SMEs, 12.5% was allocated as tax breaks in the form of corporate tax cuts and grants, and the balance was used to support households through personal income tax rebates and the securing of jobs by subsiding wages. Actual spending accounted for 62.5% of Singapore's stimulus package. Singapore was able to draw from its US$200.2bn (SGD300bn) reserves to meet its record US$13.7bn (SGD20.5bn) spending
Taiwan	Nov 2008	#6 US$15.1bn (TWD500bn)	Stimulus package	Including a shopping voucher program, the launch of public construction projects, urban renewal plans, and incentives to encourage private investment and industrial upgrading

Sources: (marked #1–#6 in the third column)

1. Abidin MZ (July, 2010) Fiscal policy coordination in Asia: East Asian infrastructure investment fund. Asian Development Bank Institute Working Paper 232. Retrieved from http://www.adbi.org/working-paper/2010/07/30/3973.fiscal.policy.coordination.asia/east.asian.response.measures/
2. Heilmann S, Schmidt D (2010) China country report. In Bertelsmann Stiftung (ed) Managing the crisis. A comparative assessment of economic governance in 14 economies. Bertelsmann Stiftung, Gütersloh. Retrieved from http://www.bertelsmann-transformation-index.de/pdf/China.pdf
3. OECD (June, 2009) Policy responses to the economic crisis: investing in innovation for long-term growth. Retrieved from http://www.oecd.org/dataoecd/59/45/42983414.pdf
4. Hong Kong Trader (September 12, 2008) Hong Kong unveils economic stimulus package. Retrieved from http://www.hktrader.net/200812/dn/2008dec09news1.htm
5. China Briefing (February 27, 2009) Hong Kong unveils stimulus package. Retrieved from http://www.china-briefing.com/news/2009/02/27/hong-kong-unveils-stimulus-package.html
6. The China Post (November 22, 2008) Premier unveils NT$500 billion stimulus package. Retrieved from http://www.chinapost.com.tw/taiwan/t-business/2008/11/22/184287/Premier-unveils.htm

Appendix 2 Important meetings held by world leaders to address the 2008 global financial crisis

Date	Place	Participants
June 18–19, 2012	Los Cabos, Mexico	G20
June 17, 2012	Los Cabos, Mexico	G7 finance ministers
March 1–2, 2012	Brussels	European Council
January 30, 2012	Brussels	Informal meeting of members of the European Council
December 8–9, 2011	Brussels	European Council
November 12–13, 2011	Honolulu, USA	19th APEC Economic Leaders' Meeting
November 3–4, 2011	Cannes, France	G20
October 23, 2011	Brussels	European Council
September 10, 2011	Marseille	G8 Deauville Partnership finance ministers
September 9–10, 2011	Marseille	G7 finance ministers and central bank governors
July 21, 2011	Brussels	Meeting of heads of state or government of the Euro area
June 23–24, 2011	Brussels	European Council
March 24–25, 2011	Brussels	European Council
March 17, 2011	Japan	G7 finance ministers telephone conference
March 11, 2011	Brussels	Extraordinary European Council
February 4, 2011	Brussels	European Council
December 16–17, 2010	Brussels	European Council
November 21, 2010	Ireland	G7 finance ministers
November 13–14, 2010	Yokohama, Japan	18th APEC Economic Leaders' Meeting
November 11–12, 2010	Seoul, Korea	G20
October 28–29, 2010	Brussels	European Council
October 8, 2010	Washington DC	G7 finance ministers and central bank governors
September 16, 2010	Brussels	European Council
June 26–27, 2010	Toronto, Canada	G20
June 17, 2010	Brussels	European Council
March 25–26, 2010	Brussels	European Council
February 11, 2010	Brussels	Informal meeting of heads of state or government of the European Union
February 5–6, 2010	Iqaluit, Nunavut, Canada	G7 finance ministers and central bank governors
December 10–11, 2009	Brussels	European Council
November 19, 2009	Brussels	Informal meeting of heads of state or government of the European Union

(continued)

(continued)

Date	Place	Participants
November 14–15, 2009	Singapore	17th APEC Economic Leaders' Meeting
October 29–30, 2009	Brussels	European Council
October 3, 2009	Istanbul, Turkey	G7 finance ministers and central bank governors
September 24–25, 2009	Pittsburgh, USA	G20
September 17, 2009	Brussels	Informal meeting of heads of state or government of the European Union
June 18–19, 2009	Brussels	European Council
June 12–13, 2009	Lecce, Italy	G8 finance ministers
April 24, 2009	Washington DC	G7 finance ministers and central bank governors
April 2, 2009	G20 London Summit United Kingdom	G20
March 19–20, 2009	Brussels	European Council
February 2009	Rome	G7 finance ministers and central bank governors
December 11–12, 2008	Brussels	European Council
November 2008	G20 APEC Peru	G20
November 22–23, 2008	Lima, Peru	16th APEC Economic Leaders' Meeting
November 14–15, 2008	Washington D.C., USA	G7
October 15–16, 2008	Brussels	European Council
October 10, 2008	Washington, DC	G7 Finance Ministers
September 1, 2008	Brussels	Extraordinary European Council
July 7–9, 2008	Tokyo, Japan	34th G8 summit
June 19–20, 2008	Brussels	European Council
June 13–14, 2008	Osaka	G8 finance ministers meeting
March 13–14, 2008	Brussels	European Council
February 9, 2008	Tokyo	G7 finance ministers and central bank governors meeting

For further information:
http://g20mexico.org/en/g20/previous-leaders-summits
http://video.consilium.europa.eu/ecofin/
http://www.g8.utoronto.ca/finance/index.htm
http://www.european-council.europa.eu/council-meetings/conclusions
http://www.apec.org/

Appendix 3 Indicators in each type of capital

Human capital index	Market capital index
1. Skilled labor*	1. Corporate tax*
2. Employee training*	2. Cross-border venture*
3. Literacy rate	3. Openness of culture*
4. Higher education enrollment	4. Globalization*
5. Pupil-teacher ratio	5. Transparency*
6. Internet subscribers	6. Image of country*
7. Public expenditure on education	7. Exports of goods
Process capital index	Renewal capital index
1. Business competition environment*	1. Business R&D spending
2. Government efficiency*	2. Basic research*
3. Intellectual property rights protection*	3. R&D spending/GDP
4. Capital availability*	4. R&D researchers
5. Computers in use per capita	5. Cooperation between universities and enterprises*
6. Convenience of establishing new firms*	6. Scientific articles
7. Mobile phone subscribers	7. Patents per capita (USPTO + EPO)

Remarks:
1. Financial capital is the logarithm of GDP per capita adjusted by purchasing power parity
2. Indicators marked with an asterisk are rated qualitatively using a scale of 1–10
3. Indicators with objective numbers are transformed to scores between 1 and 10 by calculating the ratio to the highest number in that indicator and then multiplying by 10
4. Readers can refer to the book *National Intellectual Capital: A Comparison of 40 Countries* for model validation details (www.nic40.org)

Appendix 4 Definition of the 29 indicators

Indicators	Definition
Human capital	
Skilled labor*	Whether skilled labor is readily available
Employee training*	Whether employee training is a high propriety in companies
Literacy rate	Adult (over 15 years) literacy rate as a percentage of population
Higher education enrollment	Percentage of population that has attained at least tertiary education
Pupil-teacher ratio	Ratio of teaching staff to students
Internet subscribers	Number of internet users per 1000 people
Public expenditure on education	Total public expenditure on education (percentage of GDP)
Market capital	
Corporate tax encouragement*	Whether corporate taxes encourage entrepreneurial activity
Cross-border venture*	Whether international transactions can be freely negotiated with foreign partners

(continued)

(continued)

Indicators	Definition
Openness to foreign culture*	Whether the national culture is open to foreign culture ideas
Attitudes toward globalization*	Whether attitudes toward globalization are generally positive in a given society
Transparency*	Whether transparency of government policy is satisfactory
Country image*	Whether the image abroad of a given country encourages business development
Exports of goods	Exports of goods (percentage of GDP)
Process capital	
Business competition environment*	Whether competition legislation is efficient in preventing unfair competition
Government efficiency*	Whether government bureaucracy hinders business activity
Intellectual property rights protection*	Whether intellectual property rights are adequately enforced
Capital availability*	Whether cost of capital encourages business development
Computers in use per capita	Number of computers per 1000 people
Convenience of establishing new firms*	Whether creation of firms is supported by legislation
Mobile phone subscribers	Number of subscribers per 1000 inhabitants
Renewal capital	
Business R&D spending	Business expenditure on R&D (per capita)
Basic research*	Whether basic research enhances long-term economic development
R&D spending/GDP	Total expenditure on R&D (percentage of GDP)
R&D researchers	Total R&D personnel nationwide per capita (full-time work equivalent per 1000 people)
Cooperation between universities and enterprises*	Whether knowledge transfer is highly developed between universities and companies
Scientific articles	Scientific articles published by origin of author (per capita)
Patents per capita (USPTO + EPO)	USPTO and EPO total patents granted (per capita)
Financial capital	
GDP per capita (PPP) (from IMF)	Gross domestic product based on purchasing power parity (PPP) per capita

Appendix 5 48 countries by cluster and by continent

This booklet series covers a total of 48 countries in 11 country clusters. China, Brazil, and South Africa are grouped twice to maintain cluster completeness. China is in both Greater China region and BRICKS region, Brazil is in both BRICKS region and Latin American region, and South Africa is in both BRICKS region and in Cluster 5

Eleven country clusters:

1. Portugal, Italy, Greece, Spain
2. Greater China region (China, Hong Kong, Singapore, Taiwan)
3. Brazil, Russia, India, China, Korea, South Africa (BRICKS)
4. Indonesia, Malaysia, the Philippines, Thailand
5. Israel, Jordan, Turkey, South Africa
6. Bulgaria, Czech Republic, Hungary, Romania, Poland
7. Austria, Belgium, the Netherlands, Switzerland
8. France, Germany, Ireland, United Kingdom
9. Argentina, Brazil, Chile, Colombia, Mexico, Venezuela
10. Australia, Canada, Japan, New Zealand, USA
11. Denmark, Finland, Iceland, Norway, Sweden

48 countries by continent:

Europe—24 countries

Nordic Europe—Denmark, Finland, Iceland, Norway, Sweden

Western Europe—Austria, Belgium, France, Germany, Ireland, the Netherlands, Switzerland, the UK

Southern Europe—Greece, Italy, Portugal, Spain

Eastern Europe—Russia

East Central Europe—Bulgaria, Czech Republic, Hungary, Poland, Romania, Turkey

Americas—8 countries

North America—Canada, USA

Latin America—Argentina, Brazil, Chile, Colombia, Mexico, Venezuela

Australia—2 countries

Australia and New Zealand

Asia—13 countries

East Asia—China, Hong Kong, Japan, Korea, Taiwan

West Asia—Israel, Jordan

Southeast Asia—Indonesia, Malaysia, the Philippines, Singapore, Thailand

South Asia—India

Africa—1 country

South Africa

Appendix 6 National intellectual capital scores and ranking for 48 countries (2005–2010)

2005–2010	Human capital		Market capital		Process capital		Renewal capital		Financial capital		NIC	
Mean	6.420		5.409		5.345		3.658		9.115		29.946	
SD	1.217		1.085		1.454		2.056		0.690		5.840	
Country	Score	Ranking	Score	Ranking	Score	Ranking	Score	Ranking	Score	Ranking	Score	Ranking
Argentina	5.370	36	3.340	47	3.038	47	1.413	45	8.755	34	21.916	45
Australia	7.205	15	6.199	12	6.938	9	4.927	16	9.695	11	34.964	13
Austria	7.229	14	6.203	11	6.982	8	5.296	12	9.708	10	35.418	11
Belgium	7.666	8	5.519	23	5.924	21	4.706	17	9.635	14	33.450	19
Brazil	4.757	44	4.589	38	3.076	46	1.753	36	8.481	40	22.656	42
Bulgaria	5.493	34	4.929	31	4.054	36	1.589	41	8.646	37	24.710	35
Canada	7.855	6	6.137	13	6.464	15	5.064	14	9.710	9	35.229	12
Chile	5.414	35	6.663	5	5.163	28	1.710	37	8.784	32	27.734	30
China	4.942	39	5.184	28	3.746	40	2.099	33	7.967	44	23.938	40
Colombia	4.941	40	4.677	36	3.951	37	1.435	44	8.348	42	23.352	41
Czech Republic	5.995	30	5.720	18	5.378	25	2.905	24	9.271	27	29.269	25
Denmark	8.545	1	6.662	6	7.642	1	6.587	6	9.665	13	39.101	3
Finland	7.601	9	6.232	10	7.399	3	7.755	3	9.606	17	38.593	4
France	6.956	20	4.141	45	5.730	22	4.500	20	9.585	18	30.912	23
Germany	6.761	23	5.630	19	6.489	14	5.796	9	9.612	16	34.287	17
Greece	5.986	31	4.216	43	4.113	35	2.168	32	9.440	23	25.923	31
Hong Kong	6.564	25	8.090	2	7.209	5	3.277	23	9.794	4	34.934	14
Hungary	6.674	24	4.836	32	5.095	29	2.301	29	9.046	29	27.952	29
Iceland	8.357	3	5.614	21	6.561	12	5.931	8	9.713	8	36.154	8
India	3.926	48	5.620	20	3.556	42	1.954	34	7.285	48	22.340	44
Indonesia	4.062	47	4.832	33	3.257	44	1.519	42	7.594	46	21.263	47
Ireland	7.145	16	6.816	3	6.711	11	4.101	21	9.768	5	34.541	16

(continued)

(continued)

2005–2010	Human capital		Market capital		Process capital		Renewal capital		Financial capital		NIC	
Israel	8.040	4	5.954	16	6.454	17	7.091	4	9.414	24	36.953	6
Italy	6.471	28	4.155	44	5.184	27	2.745	26	9.479	22	28.035	28
Japan	7.581	10	4.820	34	6.209	18	6.473	7	9.576	19	34.659	15
Jordan	5.345	37	5.341	26	4.411	32	2.212	30	7.865	45	25.174	33
Korea	6.964	19	5.123	29	5.282	26	4.595	19	9.383	25	31.348	22
Malaysia	6.538	26	6.410	9	5.567	23	2.186	31	8.740	35	29.442	24
Mexico	4.871	42	4.491	39	3.226	45	1.237	48	8.781	33	22.606	43
Netherlands	7.312	12	6.761	4	6.879	10	5.163	13	9.739	7	35.855	10
New Zealand	7.123	18	5.999	15	6.460	16	3.701	22	9.380	26	32.662	21
Norway	7.936	5	6.055	14	7.008	7	4.961	15	9.987	1	35.947	9
Philippines	4.772	43	4.464	40	3.317	43	1.268	47	7.481	47	21.302	46
Poland	6.393	29	4.246	41	3.895	38	1.819	35	8.945	30	25.299	32
Portugal	6.882	22	4.801	35	5.470	24	2.467	27	9.224	28	28.844	26
Romania	5.829	32	4.219	42	3.806	39	1.685	38	8.594	39	24.133	39
Russia	5.530	33	3.860	46	3.588	41	2.349	28	8.822	31	24.150	38
Singapore	7.144	17	8.306	1	7.506	2	5.416	11	9.961	2	38.334	5
South Africa	4.708	45	4.976	30	4.472	31	1.677	39	8.472	41	24.306	37
Spain	6.524	27	4.649	37	5.033	30	2.787	25	9.484	21	28.476	27
Sweden	8.385	2	6.563	7	7.174	6	7.780	2	9.673	12	39.575	1
Switzerland	7.575	11	6.481	8	7.374	4	7.988	1	9.753	6	39.172	2
Taiwan	7.244	13	5.823	17	6.026	20	5.425	10	9.526	20	34.044	18
Thailand	5.310	38	5.524	22	4.365	33	1.436	43	8.304	43	24.938	34
Turkey	4.703	46	5.227	27	4.167	34	1.666	40	8.690	36	24.453	36
United Kingdom	6.939	21	5.440	24	6.090	19	4.697	18	9.626	15	32.793	20
USA	7.695	7	5.348	25	6.558	13	6.715	5	9.886	3	36.202	7
Venezuela	4.884	41	2.761	48	2.545	48	1.277	46	8.625	38	20.092	48

Appendix 7 Country profile: additional statistics

1. International reserves in millions of US$

Country	2005	2006	2007	2008	2009	2010
China	663,190	877,637	1,204,035	1,684,280	1,956,830	2,463,547
Hong Kong	122,378	125,809	135,326	160,712	186,226	258,752
Singapore	112,752	121,766	137,728	177,462	166,251	196,361
Taiwan	251,135	257,051	267,485	286,860	300,122	355,035

Source: Global Finance http://www.gfmag.com/

2. GDP per capita, ppp (current international dollar)

Country	2005	2006	2007	2008	2009	2010
China	4,103	4,749	5,554	6,189	6,786	7,519
Hong Kong	35,550	38,877	42,313	43,881	42,757	45,736
Singapore	43,976	47,346	50,269	50,582	49,764	56,522
Taiwan	26,657	28,888	31,405	32,215	31,770	35,227

Source: World Economic Outlook Database http://www.imf.org/external/ns/cs.aspx?id=28

3. GDP (PPP): share of world total_%

Country	2005	2006	2007	2008	2009	2010
China	9.46	10.14	11.00	11.74	12.95	13.61
Hong Kong	0.43	0.44	0.44	0.44	0.43	0.44
Singapore	0.34	0.35	0.37	0.36	0.36	0.39
Taiwan	1.07	1.07	1.08	1.06	1.05	1.11

Source: World Economic Outlook Database http://www.imf.org/external/ns/cs.aspx?id=28

4. Gross domestic expenditure on R&D as a % of GDP

Country	2005	2006	2007	2008	2009	2010
China	1.23	1.33	1.42	1.49	1.54	1.7
Hong Kong	0.69	0.74	0.79	0.81	0.73	0.79
Singapore	2.24	2.36	2.31	2.27	2.68	2.27
Taiwan	2.42	2.52	2.58	2.62	2.77	2.94

Source: IMD World Competitiveness Yearbook 2006–2011

Appendix 8 Export and import dependency within the Greater China economies

		China	Hong Kong	Singapore	Taiwan
China	Export		14.1%		
	Import				
Hong Kong	Export	52.4%			
	Import	44.9%			7.5%
Singapore	Export	10.4%	11%		
	Import	10.4%			5.9%
Taiwan	Export	28.1%	13.8%	4.4%	
	Import	14.2%			

Source: The U.S. Intelligence Agency, the World Factbook March 2012 data

Glossary

12th Five-Year Plan China's ambitious 12th Five-Year Plan
APEC Asia-Pacific Economic Cooperation
ASEAN Association of Southeast Asian countries (including Brunei, Burma, Cambodia, Indonesia, Laos, Malaysia, the Philippines, Singapore, Thailand, and Vietnam), China, and India.
Basel II **Basel II** is the second of the Basel Accords, which are recommendations on banking laws and regulations issued by the Basel Committee on Banking Supervision. Basel II, initially published in June 2004, was intended to create an international standard for banking regulators to control how much capital banks need to put aside to guard against the types of financial and operational risks banks face.
BBC British Broadcasting Corporation
BRICKS Brazil, Russia, India, China, Korea, South Africa
CIA The U.S. Central Intelligence Agency
CNY Chinese Yuan—formal name of Chinese currency, also known as RMB
CPI Consumer Price Inflation
ECFA Economic Cooperation Framework Agreement signed by China and Taiwan in June 2010 to strengthen the trade relationship between the two economies, by lowering tariffs on a number of goods and by reducing market access barriers for services
EU European Union
FC Financial Capital
FDI Foreign Direct Investment
GCI Global Competitiveness Index (published by World Economic Forum)
GDP Gross Domestic Product
GDP per capita (ppp) Gross domestic product per capita adjusted by purchasing power parity
HC Human Capital
IMD International Institute for Management Development, Switzerland

C.Y.-Y. Lin et al., *National Intellectual Capital and the Financial Crisis in China, Hong Kong, Singapore, and Taiwan*, SpringerBriefs in Economics 8,
DOI 10.1007/978-1-4614-5984-2, © The Author(s) 2013

IMF International Monetary Fund

IP Intellectual Property

IPO Initial Public Offering

Long-term NIC Long-term oriented national intellectual capital, represented by human capital and renewal capital

MAS Monetary Authority of Singapore

MC Market Capital

NIC National Intellectual Capital

OECD Organization for Economic Cooperation and Development

OEM Original Equipment Manufacturing

PC Process Capital

Ppp Purchasing Power Parity

R&D Research and Development

RC Renewal Capital

RMB Currency in China

SAR Special Administrative Region, Hong Kong is the SAR of the People's Republic of China

Short-term NIC Short-term oriented national intellectual capital, represented by market capital and process capital

SMEs Small and Medium-Sized Enterprises

Three time periods For the convenience of comparing ranking changes, the six years data were separated into three time periods, namely, 2005–2006, 2007–2008, and 2009–2010. Two years each represents "before, during, and after" the 2008 global financial crisis. Even though Euro zone is still in sovereign debt trouble as of the mid-2012, the time partition is based on the financial crisis progression in the epicenter. The signs of financial trouble appeared in 2007, and the crisis was declared over in the third quarter of 2009.

UNCTAD United Nations Conference on Trade and Development

USPTO United States Patent and Trademark Office

VAT Value-Added Tax

WTO World Trade Organization

References

Abidin MZ (2010) Fiscal policy coordination in Asia: East Asian infrastructure investment fund. Asian Development Bank Institute Working paper Series #232. Retrieved from http://www.adbi.org/files/2010.07.30.wp232.fiscal.policy.coordination.asia.pdf

Adam S (November 14, 2010) China to exceed U.S. by 2020, Standard Chartered says. Bloomberg

Altman RC (January 2, 2009) The great crash, 2008: a geopolitical setback for the west, Foreign Affair. Retrieved from http://www.foreignaffairs.com/articles/63714/roger-c-altman/the-great-crash-2008

Asher MG (December 31, 2010) Singapore: economy bounces back, but…? East Asia Forum. Retrieved from http://www.eastasiaforum.org/2010/12/31/singapore-economy-bounces-back-but/

Barta P, Holmes S (February 2, 2010) Singapore panel issues subdued economic forecast. The Wall Street Journal

Bradsher K (September 23, 2011) China, driver of world economy, may be slowing. The New York Times. Retrieved from http://www.nytimes.com/2011/09/24/business/global/chinas-economic-engine-shows-signs-of-slowing.html?pagewanted=all

Bryant L (September 8, 2011) Growth expected to slow in world's largest economies. Voice of America News. Retrieved from http://www.voanews.com/english/news/europe/OECD--Economic-Slowdown-Likely-to-Continue-129458543.html

Business Anti-corruption Portal (2012) China country profile. Retrieved from http://www.business-anti-corruption.com/country-profiles/east-asia-the-pacific/china/snapshot/ in May 2012

Central Intelligence Agency (CIA) (2012) The world factbook. U.S.A. Central Intelligence Agency. Retrieved from https://www.cia.gov/library/publications/the-world-factbook/geos/ch.html

Chan S (January 25, 2011) Financial crisis was avoidable, inquiry finds. The New York Times. Retrieved from http://www.nytimes.com/2011/01/26/business/economy/26inquiry.html?_r=2&scp=1&sq=financial%20crisis%20was%20avoidable,%20inquiry%20finds&st=cse

Cheng YH, Kao YL (January 7, 2010) Taiwan's economy forecast to grow 4.81% in 2010. The China Post. Retrieved from http://www.chinapost.com.tw/business/asia/b-taiwan/2010/01/26/242417/Taiwans-economy.htm

China Economic Net (December 9, 2011) Hong Kong to grow 4 pct in 2012: IMF. Retrieved from http://en.ce.cn/Business/Macro-economic/201112/09/t20111209_22907232.shtml

Cooke MT (April 7, 2012) Taiwan's economy: recovery with Chinese characteristics. The Brookings Institution. Retrieved from http://www.brookings.edu/opinions/2009/11_taiwan_economy_cooke.aspx

Debnath A (January 19, 2012) Asian economies to slow in 2012 but avoid slump. The Reuters. Retrieved from http://in.reuters.com/article/2012/01/19/asia-economy-poll-idINDEE80I05S20120119

C.Y.-Y. Lin et al., *National Intellectual Capital and the Financial Crisis in China, Hong Kong, Singapore, and Taiwan*, SpringerBriefs in Economics 8,
DOI 10.1007/978-1-4614-5984-2, © The Author(s) 2013

Deutsche Bank Research (June 19, 2009) Hong Kong's banks during the financial crisis. Retrieved from (http://www.dbresearch.de/PROD/DBR_INTERNET_EN-PROD/PROD0000000000242795.pdf)
Devan J, Negri S, Woetzel JR (2008) Meeting the challenges of China's growing cities. Mckinsey Quarterly. Retrieved from https://www.mckinseyquarterly.com/Economic_Studies/Country_Reports/Meeting_the_challenges_of_Chinas_growing_cities_2152
Drysdale P (October 10, 2010) Asia's global responsibilities: delivering through global and regional arrangements. East Asia Forum. Retrieved from http://www.eastasiaforum.org/2010/10/10/asias-global-responsibilities-delivering-through-global-and-regional-arrangements/
Economic Forum (December 6, 2010) Hong Kong 2011 economic outlook. Economic Forum – Global Economic Analysis & Reports. Retrieved from http://www.hktdc.com/info/mi/a/ef/en/1X078D1W/1/Economic-Forum/Hong-Kong-2011-Economic-Outlook-Living-With-Higher-Prices.htm
Economic Forum (March 15, 2011) Hong Kong consumer confidence experiences slight dip after steady increment for eight quarters. Economic Forum – Global Economic Analysis & Reports. Retrieved from http://www.hktdc.com/info/mi/a/ef/en/1X07EDQ7/1/Economic-Forum/Hong-Kong-Consumer-Confidence-Experiences-Slight-Dip-After-Steady-Increment-For-Eight-Quarters-Br-Consumer-Confidence-Ended-2010-At-99-Points.htm
Economy Watch (June 30, 2010a) The Chinese economy. Economy Watch. Retrieved from http://www.economywatch.com/world_economy/china/
Economy Watch (March 18, 2010b) Singapore economy. Economy Watch. Retrieved from http://www.economywatch.com/world_economy/singapore/?page=full
Edvinsson L, Malone MS (1997) Intellectual capital: realizing your company's true value by finding its hidden brainpower. HarperBusiness, New York
Europa (November 26, 2008) The Commission launches a major Recovery Plan for growth and jobs, to boost demand and restore confidence in the European economy. Retrieved from http://europa.eu/rapid/pressReleasesAction.do?reference=IP/08/1771&format=HTML&aged=0&language=EN&guiLanguage=en
Fackler M (October 23, 2008) Financial crisis spreads to emerging nations. The New York Times. Retrieved from http://www.nytimes.com/2008/10/24/business/worldbusiness/24won.html
Fleet JD (January 21, 2010) Can China manage its economy? The Diplomat. Retrieved from http://the-diplomat.com/2010/01/21/can-china-manage-its-economy/
Fong PE (January 2, 2009) Singapore: positioned to weather the global shock. East Asia Forum. Retrieved from http://www.eastasiaforum.org/2009/01/02/singapore-positioned-to-weather-the-global-shock/
Global Edge (2012) Taiwan: economy. Michigan State University. Retrieved from http://globaledge.msu.edu/Countries/Taiwan/economy in June 2012
Global Finance (April 2011) Taiwan roundtable: economy & business, moderated by Joseph Giarraputo. Global Finance. Retrieved from http://www.gfmag.com/archives/137-april-2011/11181-roundtable-taiwans-economic-and-business-outlook.html#axzz1rOjkLER0
Goh CT (August 3, 2009) 10 challenges for Singapore. Asiaone News. Retrieved from http://www.asiaone.com/News/AsiaOne%2BNews/Singapore/Story/A1Story20090803-158678.html
Heilmann S, Schmidt D (2010) China country report. In Bertelsmann Stiftung (ed) Managing the crisis. A comparative assessment of economic governance in 14 economies. Bertelsmann Stiftung, Gütersloh
Herd R, Koen V, Noord P (2011) China's emergence as a market economy: achievements and challenges. OECD contribution to the China Development Forum 20–21 March 2011. Beijing.
Heritage Foundation (2012) Index of economic freedom. Retrieved from http://www.heritage.org/index/ranking
Hong Kong Government (2009, 4th Q) Outlook for the Hong Kong economy in 2010. Retrieved from http://www.hkeconomy.gov.hk/en/pdf/09q4_outlook.pdf
International Monetary Fund (IMF) (September 26, 2011) China economic outlook. IMF Resident Representative Office, China. Retrieved from http://www.imf.org/external/country/CHN/rr/2011/092611.pdf

International Monetary Fund (IMF) (January 24, 2012) World economic outlook update: global recovery stalls, downside risks intensify, International Monetary Fund. Retrieved from http://www.imf.org/external/pubs/ft/weo/2012/update/01/index.htm

Kehoe T (2010) The current financial crisis in Spain: what should we learn from the great depressions of the twentieth century? University of Minnesota and Federal Reserve Bank of Minneapolis, 18th Barcelona Lecture. Barcelona G.S.E. and Banc Sabadell, Feb 2010

Kesavapany K (January 5, 2010) Singapore economy on the way up again. East Asia Forum. Retrieved from http://www.eastasiaforum.org/2010/01/05/singapore-economy-on-the-way-up-again/

Khemka S, Chou S, Fonseca O (October 29, 2008) Feeling the pain: how the financial crisis is affecting Brazil, Russia, India and China. Knowledge@Wharton podcast, University of Pennsylvania. Retrieved from http://knowledge.wharton.upenn.edu/article.cfm?articleid=2079

Kuroda H (January 26, 2012) Asia to power world economy in 2012: ADB chief. The Economic Times. Retrieved from http://economictimes.indiatimes.com/news/economy/foreign-trade/articlelist/1200949414.cms

Lau SK (August 29, 2011) Hong Kong Government to tackle livelihood & economic issues. Retrieved from http://vimeo.com/28375051

Leung CY (January 31, 2008) Hong Kong Economic problem – marginalization. Retrieved from http://hknews.hksyu.edu/index.php/Hong_Kong_Economic_problem_%E2%80%93_marginalization

Levin C, Coburn T (April 13, 2011) Wall Street and the financial crisis: anatomy of a financial collapse. United States Senate. Retrieved from http://www.levin.senate.gov/imo/media/doc/supporting/2011/PSI_WallStreetCrisis_041311.pdf

Lin CYY (2011) Leveraging intellectual capital for future competitiveness in Finland. Presentation at Tekes. Helsinki, 10 Sept 2011

Lin CYY, Edvinsson L (2011) National intellectual capital: a comparison of 40 countries. Springer Publishing Co, New York

Liu HC, Wu S (October 14, 2011) Taiwan to be free of 'double-dip recession': CEPD chief. Taiwan News. Retrieved from http://www.taiwannews.com.tw/etn/news_content.php?id=1735689

Malhotra Y (2003) Managing and measuring knowledge assets in the public sector. Working Paper, Syracuse University

Ministry of Trade and Industry Singapore – MTI (2011) MTI forecasts growth of 5.0% in 2011 and 1.0–3.0% in 2012. MTI Singapore Press Release. Retrieved from http://online.wsj.com/article/SB10001424052970204720204577132040862931660.html

Morrison WM (2011) China's economic conditions. Congressional Research Service report for Congress, 7–5700. Retrieved from http://www.fas.org/sgp/crs/row/RL33534.pdf

Nanto DK (October 2, 2009) The global financial crisis: analysis and policy implications. CRS Report for Congress, Congressional Research Service 7–5700, RL34742

National Statistics, Taiwan (2012) Taiwan's economic performance in 2011 Q2 and outlook for 2011–2012. Retrieved from http://eng.stat.gov.tw/ct.asp?xItem=29664&ctNode=3274 in June 2012

Norris F (March 10, 2011) Crisis is over, but where's the fix? The New York Times. Retrieved from http://www.nytimes.com/2011/03/11/business/economy/11norris.html?pagewanted=1

OECD (March 2008) China's outward foreign direct investment. OECD Investment. Retrieved from http://www.oecd.org/dataoecd/28/10/40283257.pdf

OECD (2010) 2010 Economic Surveys Portugal, OECD. Retrieved from http://www.oecd.org/dataoecd/34/20/46062172.pdf

Pierson D (October 15, 2011) Inflation remains a worry in China. Los Angeles Times. Retrieved from http://articles.latimes.com/2011/oct/15/business/la-fi-china-inflation-20111015

QFinance (2011) Taiwan – Economy. Retrieved from http://www.qfinance.com/country-profiles/taiwan

Reavis C (July 22, 2009) The global financial crisis of 2008–2009: the role of greed, fear and oligarchs. MIT Sloan Management Case #09-093

Reinhart CM, Rogoff K (2009) This time is different: eight centuries of financial folly. Princeton University Press

Reuters (September 11, 2011) Beijing to fund VCs to aid local hi-tech. Retrieved from Reuters http://www.reuters.com/article/2011/09/11/us-china-vc-idUSTRE78A0U620110911

Reuters (March 16, 2012) Singapore economy to grow 2.5 pct in 2012 – MAS Poll. Retrieved from Reuters http://www.reuters.com/article/2012/03/16/singapore-economy-survey-idUSS7E7L401D20120316

Schwab K (ed) (2011) The global competitiveness report 2011–2012. World Economic Forum, Geneva

Shannon K (July 15, 2010) Singapore's economy leads Asia's rebound with record-breaking 2010 growth. Money Morning. Retrieved from http://moneymorning.com/2010/07/15/singapores-economy/

Thangavelu S (January 5, 2009) Riding the global economic crisis in Singapore. East Asia Forum. Retrieved from http://www.eastasiaforum.org/2009/01/05/riding-the-global-economic-crisis-in-singapore/

Tuan C, Ng LFY (2011) Beyond the financial crisis: Hong Kong-metropolis's evolution and economic integration with China. Department of Decision Sciences and Managerial Economics, Faculty of Business Administration. The Chinese University of Hong Kong. Retrieved from http://ccas.doshisha.ac.jp/eng/Econference/econfpdf/26th%20Conference%20of%20the%20American%20Committee%20for%20Asian%20Economic%20Studies/TuanNg.pdf

US News (July 8, 2008) China's economy is on track to surpass the U.S. by 2035. US News. Retrieved from http://www.usnews.com/news/articles/2008/07/08/chinas-economy-is-on-track-to-surpass-the-us-by-2035

Venkat PR (December 31, 2011) Singapore economy expanded 4.8% in 2011. The Wall Street Journal. Retrieved from http://online.wsj.com/article/SB10001424052970204720204577132040862931660.html

World Bank (1998) World development report: knowledge for development. Oxford University Press

World Trade Organization (WTO) (2012) Member information: China and the WTO. World Trade Organization. Retrieved from http://www.wto.org/english/thewto_e/countries_e/china_e.htm in July 2012

Xu XP (August 4, 2010) A deal that will shape Taiwan's economic future in Asia. East Asia Forum. Retrieved from http://www.eastasiaforum.org/2010/08/04/a-deal-that-will-shape-taiwans-economic-future-in-asia/

Yang Z, Tong SY (April 14, 2009) Hong Kong's economy in the financial crisis. EAI Background Brief No. 448. Retrieved from http://www.eai.nus.edu.sg/BB448.pdf

Author Index

C.Y.-Y. Lin et al., *National Intellectual Capital and the Financial Crisis in China, Hong Kong, Singapore, and Taiwan*, SpringerBriefs in Economics 8,
DOI 10.1007/978-1-4614-5984-2, © The Author(s) 2013

Subject Index

C.Y.-Y. Lin et al., *National Intellectual Capital and the Financial Crisis in China, Hong Kong, Singapore, and Taiwan*, SpringerBriefs in Economics 8,
DOI 10.1007/978-1-4614-5984-2, © The Author(s) 2013

MIX
Papier aus verantwortungsvollen Quellen
Paper from responsible sources
FSC® C105338

If you have any concerns about our products,
you can contact us on
ProductSafety@springernature.com

In case Publisher is established outside the EU,
the EU authorized representative is:
Springer Nature Customer Service Center GmbH
Europaplatz 3, 69115 Heidelberg, Germany

Printed by Libri Plureos GmbH
in Hamburg, Germany